What Mothers Say about Special Education

Palgrave Studies in Urban Education

Series Editors: Alan R. Sadovnik and Susan F. Semel

Reforming Boston Schools, 1930–2006: Overcoming Corruption and Racial Segregation
By Joseph Marr Cronin (April 2008)

What Mothers Say about Special Education: From the 1960s to the Present
By Jan W. Valle (March 2009)

Charter Schools: From Reform Imagery to Reform Reality
By Jeanne M. Powers (forthcoming)

The History of "Zero Tolerance" in American Public Schooling
By Judith Kafka (forthcoming)

What Mothers Say about Special Education

From the 1960s to the Present

Jan W. Valle

palgrave
macmillan

KH

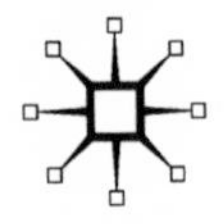

WHAT MOTHERS SAY ABOUT SPECIAL EDUCATION

First published in 2009 by
PALGRAVE MACMILLAN®
in the United States—a division of St. Martin's Press LLC,
175 Fifth Avenue, New York, NY 10010.

Where this book is distributed in the UK, Europe and the rest of the world, this is by Palgrave Macmillan, a division of Macmillan Publishers Limited, registered in England, company number 785998, of Houndmills, Basingstoke, Hampshire RG21 6XS.

Palgrave Macmillan is the global academic imprint of the above companies and has companies and representatives throughout the world.

Palgrave® and Macmillan® are registered trademarks in the United States, the United Kingdom, Europe and other countries.

ISBN-13: 978–0–230–60651–7
ISBN-10: 0–230–60651–2

Library of Congress Cataloging-in-Publication Data

Valle, Jan W., 1956–
What mothers say about special education : from the 1960s to the present / by Jan W. Valle.
p. cm.—(Palgrave studies in urban education)
Includes bibliographical references.
ISBN 0–230–60651–2
1. Special education—United States. 2. Learning disabled children—Education—United States—Case studies. I. Title.

LC3981.V24 2009
371.90973—dc22 2008031123

A catalogue record of the book is available from the British Library.

Design by Newgen Imaging Systems (P) Ltd., Chennai, India.

First edition: March 2009

10 9 8 7 6 5 4 3 2 1

Printed in the United States of America.

Transferred to Digital Printing in 2009

12/10/10

To my mother
Myra Weatherly
and
in memory of my children
Jessica, Zachary, and Franco

Contents

Series Editor's Foreword

Jan Valle's book looks at 15 mothers whose children have special needs, from the early 1960s through 2004, roughly 40 years. At the time her work begins, Willowbrook was still operating, warehousing youth and adults in deplorable conditions, not unlike the prisons Dorothea Dix visited over a century ago. As Valle demonstrates to us in her book, much has changed over the time of her work, yet much remains the same.

Valle writes both scholarly and movingly about the experiences of mothers of special needs children who learn, often painfully, how to negotiate the system to obtain basic social and medical services that their children are entitled to and to ensure that they be educated to the best of their abilities. Novices at the birth of their children, many become experts, cognizant of their legal rights with various degrees of success, in dealing with medical and educational professionals. What is so important about this book is that it illuminates the actual experiences of mothers over a historical period in which the laws were changing, but often not, the attitudes of the professionals.

I speak from firsthand experience. Although not one of Valle's mothers in the book, I am the mother of a special needs child, Margaret "call me Mags" Semel, born in 1969. From the beginning I knew that something was wrong when she did not develop according to what my old undergraduate psychology textbook stated as the norm. However, I will never forget the words of my pediatrician, who informed me when I raised the issue of Mags not speaking, that "even the most severely retarded children learn to talk." As it was before the passage of the Education for All Handicapped Children Act (PL 94–142), from then on, it was a constant struggle to find speech therapists, psychologists, educators, and of course, schools that would be willing to take Mags and work with her. Along the way I was told by medical professionals that I was "hostile to the medication" they prescribed, even though I knew that it made her more hyperactive, not less as it was supposed to, and that because she did not follow instructions, she might be deaf. Only a few of the medical professionals who I encountered over 39 years actually asked me what I thought.

Thus Valle speaks for all mothers of special needs children in her book. It is history as seen through a particular group of mothers—women who are often overlooked by special educators and healthcare professionals and who work tirelessly to help their children achieve their potentials often against great odds. Her book makes an important contribution to the literature on special needs and disability studies, but it also personalizes a very special group of women whose stories needed to be told.

This book comes at an important time. With debates among academics over the neurological versus social determinants of disabilities—neuroscientists arguing that disabilities are organically and biologically based and disability studies theorists often arguing that they are socially constructed—it is essential that the voices of families be part of the conversation. Through her interviews with mothers, Valle's book is a powerful investigation into the world of mothers and their children with disabilities and the necessity for medical professionals and school professionals to listen to them.

SUSAN F. SEMEL
New York City
August 2008

Preface

I chose special education as my college major the same year that the ninety-fourth United States Congress committed to provide a "free and appropriate education for all handicapped [*sic*] children" (Education for All Handicapped Children Act [PL 94–142], 1975). As a college student during an era dubbed The Me Generation, I longed for the political activism that had defined the 1960s. It was through my studies in special education that I discovered political activism—alive and well—in the parents of children with disabilities and their advocates whose efforts redefined a nation's response to disability. Building upon the momentum of the civil rights movement, parents and advocates likewise drew upon the fourteenth amendment to challenge the separate and unequal education of children with disabilities. I clamored to join the ranks of this educational revolution that culminated in the passage of PL 94–142—the most sweeping legislation for persons with disabilities that Congress had passed in our country's history.

In the fall of 1978, the year in which states were required to have implemented PL 94–142, I began as a first-year teacher in a middle school learning disability resource room. I imagined entering an educational context that embodied the spirit of the law. Instead I met a school community—not unlike most of this era—that viewed the new law's complex requirements for institutional structures and regulations as a considerable intrusion. It was a transition marked by resentment and resistance on the part of many administrators and classroom teachers.

Even more surprising to me, I found that parents seemed disconnected to the law that promised a free and appropriate public education for their children with disabilities. So, as a young, White, middle-class teacher, I set out to cultivate meaningful partnerships with the parents of my students. I was astonished that my overtures were met with responses ranging from surprise to suspicion. From a parent perspective, uninitiated contact from school personnel rarely signaled a prelude to positive conversation. In fact, parents steeled themselves to hear a litany of offenses committed by their child, followed by the familiar demand that they work with the school to resolve the problem. The newly available special education classes represented hope for many parents that their children's problems would be handled by school personnel *at school*. Indeed, most parents viewed my initial contacts as evidence that special education would not serve as a buffer from the school intrusions they had hoped to escape.

In thinking about how to reconfigure relationships with parents, I strategized with fellow resource teacher Carol Sunderman to develop and offer a series of parent/teacher workshops. We reasoned that informal workshops held in our classroom could create a space for positive and active exchanges with and among our parents as well as provide a forum for much needed explanations about the new law. Given the reluctance of parents to engage with school personnel, we solicited the help of a few mothers with whom we had established a working relationship. These mothers agreed to contact other parents to explain the concept of the workshops, ask for topic suggestions, and generally stimulate interest in the shared project.

Six years later, attendance at our workshops had grown exponentially and taken us all down paths never anticipated. We could not have imagined the degree of positive and productive energy that would emerge from parents and teachers coming together for the benefit of their *mutual* children. Although we began with a small group, participation in the six-session series (spaced throughout the school year to foster regular interaction) continued to grow year after year. Topics for discussion shifted with each school year, depending on the interests and needs of the parents. From the beginning, our students expressed great interest in the workshops, urging their parents to attend. Before long, our students asked if *they* could attend the meetings. The subsequent engagement between students, parents, *and* teachers became one of the most powerful aspects of the program.

It is worth noting that the integration of PL 94–142 (renamed in 1990 as the Individuals with Disabilities Education Act [IDEA]) into the public school structure was, and remains, troubled. The law ensures parents of the right to be informed, the right to be knowledgeable about the actions to be taken, the right to participate, the right to challenge, and the right to appeal. The guarantee of such parental rights, however, conflicts with the assumption of professional dominance that schools have maintained throughout the history of public education (Cutler, 2000; Skrtic, 1991; Valle & Reid, 2001). Thus, it is important to situate our workshops within this context. Despite the project's overwhelming success, district personnel remained uneasy. On more than one occasion, a district-level supervisor attended our meetings to be sure that parents were not gaining information about how to sue the school district. We were stunned to learn that the exchange of knowledge between teachers and parents could be construed by anyone as "dangerous" practice. It was then that I began to formulate a vague understanding of what I would later come to recognize as the interrelationship between power and knowledge (Foucault, 1977).

Between the mid-1980s and mid-1990s, I engaged in another kind of relationship with parents of children labeled learning disabled (LD). During this period, I worked at a developmental pediatrics clinic as an educational diagnostician on an interdisciplinary evaluation team. In light of our status as a private assessment clinic, parents began to report difficulties with school personnel considering and/or responding to non-district testing and recommendations—despite a parent's right under the law to secure independent testing. Thus, my role as an educational diagnostician necessarily expanded to include parent advocacy.

As a parent advocate, I accompanied countless parents to countless school meetings. I attended meetings at every grade level in both public and private schools. I participated in meetings ranging from single teacher to those involving a cadre of professionals—teachers, administrators, district special education supervisors, speech pathologists, school psychologists, occupational therapists, and lawyers. As a former school district employee advocating from a position within the private sector, I gained a rich perspective on the dynamics of special education meetings. I attended successful meetings in which participants engaged with one another in a respectful and collaborative manner; however, more often than not, I participated in meetings where coercive persuasion masqueraded as collaboration. The parental act of requesting advocacy exposes an inherent asymmetry in positioning among participants. Given that the law guarantees parents the right to an advocate, might the inclusion of this right be a *legal* acknowledgment of the skewed balance of power toward professionals?

It is my observation that a hierarchy of knowledge characteristically exists within special education meetings. Typically, school professionals initiate, direct, and terminate the discussion. Parents often struggle to understand the legal and scientific language that circulates among professionals, searching for space to enter the conversation. Their children, described by professionals as an amalgamation of test scores, discrepancies, deficits, and limitations, may appear unrecognizable to them (Valle & Aponte, 2002). I am aware that some mothers leave school meetings only to burst into tears in the school parking lot; school personnel, on the other hand, may regard those same meetings as productive and ultimately successful.

My experiences suggest that special education discourse (language that grows out of, drives, and sustains practice) *can* alienate parents from the collaborative process guaranteed by law. Given the field's historical roots in medicine, it is significant to acknowledge that "the medical model" grounds special education practice within public schools. For example, a patient (student) presents with symptoms (educational problems).

An expert (school psychologist) performs an examination (psychoeducational assessment) to make or rule out a diagnosis (disability). If there is a disability, a prescription (individual education plan [IEP]) is written with a recommendation for a course of treatment (special education placement and individualized instruction) as well as a follow-up plan (yearly IEP review). In other words, special education meetings often function as a means to *inform* parents about test results, diagnoses, and proposed treatment. This scenario reproduces our cultural norms for a conventional doctor/patient relationship—precluding the kind of collaboration envisioned under law.

While deeply engaged in advocacy work with parents, I found myself absorbed in a personal struggle that ultimately paralleled the experiences of parents with whom I worked. Within a period of ten years, my husband and I suffered three neonatal losses (including a son born with severe disabilities) and several miscarriages. My inclusion of this personal experience is not to take attention away from the discussion at hand, but rather to highlight how parents, particularly mothers, in other circumstances are likewise positioned within a medical/scientific discourse.

Having found myself suddenly on the other side of the desk, I struggled to make sense of the rules for behaving as a Good Patient. For the most part, I learned that a Good Patient is one who is passive and uncritical about her medical care. I also learned that medical experts, more often than not, perceived my questions as challenges to their knowledge and authority. Within a medical culture, collaboration could hardly be considered the norm. My attempts at critical engagement in the process prompted experts to label me, at one time or another, as emotional, noncompliant, and uncommitted to the regimens of reproductive technology. Contributions of self-knowledge about my own body and pregnancies were dismissed as unscientific, subjective, and irrelevant. Male physicians typically regarded my husband as rational by virtue of his gender. Moreover, a dizzying array of medical terms complicated my ability to participate in my own care. Faced with inaccessible professional vocabulary, I doggedly researched the field of reproductive technology to gain access to "the code" in order to engage more fully in the process. Such alienation from my medical experience exacerbated feelings of powerlessness, loss, and grief.

As a result of negotiating a medical/scientific discourse in my own life, I gained a more intimate understanding of the challenges parents face in navigating the discourse practices of special education. Like the parents with whom I worked, I experienced firsthand the ways in which a medical/scientific discourse can position patients in a passive and compliant role. I, too, engaged with professionals, couched within medical/scientific language and practice, who routinely dismissed other ways of knowing as well as any collaboration in the process.

My concerns, then, about a medical/scientific discourse center around its unintended consequences for those positioned in the role of "patients" as well as the lack of awareness on the part of the professional community that such consequences even have occurred. This book represents my best efforts at more deeply understanding the discursive practices of special education and the accompanying material consequences for parents, specifically mothers, who navigate *our* system in the interest of their children with disabilities.

Acknowledgments

This book owes its being to the mentorship of Professor D. Kim Reid who opened the world of scholarship to me. I am most grateful to her for challenging, encouraging, and inspiring me throughout my doctoral program at Teachers College, Columbia University. From the inception of this work to its conclusion, Professor Emerita Maxine Greene contributed her wisdom and loving support. I am privileged to have worked so closely with both of these extraordinary scholars.

Very special thanks are due to the mothers who participated in my study. This book would not exist without the courage of these women to tell their stories so that others might benefit. I am deeply grateful to each mother for her investment of time and shared commitment to the project.

I wish to thank my friend and colleague Carol Sunderman with whom I began this work more than 30 years ago. This book took root in the workshop series that we created with parents of our middle school students. My collaboration with Carol remains the most meaningful partnership of my career.

I could not have brought this project to fruition without the "cheerleaders" in my life: Pamela Carter, Stefano Castri, David Connor, Tori Gilbert, Alexis Hill, Marcia James, Amy Montanez, Barbara McGinty Pedwano, Debra Pearlman, Ellen Rice, Julie Rice, Drew Valle, Donna Volpitta, Marie Wardell, Linda Ware, Jennie Weeks, and David Wilsey.

To my mother, Myra Weatherly, I owe the stamina needed to complete this book. Her daily support gave me the energy and drive to continue writing. I am grateful for her editorial skills and legacy of "writing genes" that runs throughout our family. I would also like to acknowledge the influence of my father, Dr. Owen M. Weatherly, upon this book. The memory of his scholarship, ethics, and commitment to social justice inspired me throughout its writing.

A special thanks goes to my editors at Palgrave, Julia Cohen and Erin Ivy. I have a stronger manuscript because of their hard work and investment in this project.

Finally, none of this would have been possible without the support of my husband, Paul, who moved to New York City with me in pursuit of my dream. To him, I owe everything.

Chapter One

Setting the Stage: Context and Method

This book offers a historical account of special education discourse within American public schools from the viewpoint of 15 mothers with children labeled learning disabled (LD). It is a version of history in which mothers, representing diverse generations, social classes, races, and ethnicities, describe their lived experiences within the established structures that provide a free and appropriate public education for children with disabilities. Their collective narrative spans a period of 40 years—a time frame that begins in the 1960s (before the passage of the Education for All Handicapped Children Act [PL 94–142]) and concludes shortly after the turn of the new century.

My understanding of special education as a *discourse* is grounded in the French philosopher Michel Foucault's body of work on discursive practices (1965, 1972, 1973, 1975, 1978, 1980). Although discourse or discursive practice is often interpreted as language usage or language meaning, I rely upon Foucault's meaning of discourse as a "system of rules that defines what *can be said*—what counts as natural and true within a particular discursive practice—and the instrument through which people become *positioned*, but not determined, within that discourse" (Reid & Valle, 2004, p. 466, emphasis in original). In other words, I am interested in how mothers experience the particular way that our public school system has chosen to conceptualize, speak about, and respond to school-age children with disabilities.

In that recorded history typically reflects a dominant culture's perspective, the integrated perspective of discursive practice and narrative inquiry is a useful tool for revealing a previously undocumented account of special education discourse (with a particular emphasis on the field of learning disabilities) from the perspective of mothers. Within this particular historical reconstruction, the consequences of special education discourse upon the everyday lives of mothers become clearer. My intent, then, is to pose and explore questions regarding the ethics of special education practices in light of the experiences that mothers report. By integrating discourse analysis and narrative inquiry, I am able to map discursivity and lived experiences against the official rendering of special education in hopes of generating new ways to reflect upon our past and current practices.

A History of Special Education: Whose Viewpoint Counts?

History is inarguably an integral and valued aspect of culture. Human beings study and record their pasts in meticulous detail, telling and retelling history in the search for new and undiscovered meanings. Recorded histories represent our insatiable quest for the cohesive narrative—an evolutionary tale of progress born out of serial and significant transformations within a stable chronology of cause and effect.

Histories are collected everywhere and about everything. Beyond major historical chronicles, smaller histories exist that capture and record events and transformations more closely connected with our daily lives. Such is the case for the professions of our present-day culture. Professionals, representing disciplines such as medicine, law, psychology, education, and applied technology, subscribe to a particular history of progress relative to the development of their respective professions (Skrtic, 1991). Such histories contribute to the construction of professional identity within particular disciplines.

The learning disability profession, for example, records its history in college textbooks, academic journals, and the annals of professional organizations. It is a history told and continuously retold. Professionals in the field, along with critics from allied fields, assume their positions as foot soldiers in the march of progress—demonstrating commitment (or not) to the field's official history while contributing to its ongoing construction.

Historical chronicles can be of great value and significance in the creation of meaning. It is essential to acknowledge, however, that no single history reveals absolute Truth. Brueggemann (1999) explains that

> histories are surely useful, but they can also carry a rhetoric all their own in their partiality, in their "official" disguise, in their tendencies to cast change as "progress" moving toward things always bigger and better, and, finally, in their penchant for not being a history *of* thepeople even as they are *about* the people. (p. 27, emphasis in original)

Thus, this book offers an alternative (albeit intersecting) history told by 15 mothers whose collective narrative challenges any portrayal of special education as a tidy tale of continuous progress.

I might have written an account from the perspective of people whose careers and livelihoods depend upon the category of learning disabilities. I could have looked at the language and practices of the cadre of professionals who comprise the learning disabilities service industry, such as special education teachers, special education consultants, educational diagnosticians, school psychologists, clinical psychologists, physicians, neurologists, private consultants, speech/language pathologists, tutors, special education administrators, advocates, and lawyers. But such a study, although interesting on many levels, would have reflected yet another perspective *about* the people but not *of* the people.

The people, in this case, are the school-aged population whose perspectives and experiences of special education have been mostly overlooked in the "official" history of learning disabilities—although recent research efforts reflect movement toward eliciting voices of students labeled LD (Reid & Button, 1996; Rodis, Garrod, & Boscardin, 2000; Connor, 2007). Yet it is the responsibility of parents, not students, to *negotiate* the special education system—year by year, teacher by teacher, evaluation by evaluation, meeting by meeting. In that parental participation is guaranteed under special education law, the assumption is that parents *will* engage in the process. And there are those who do—explaining, listening, justifying, bargaining, contributing, suggesting, pleading, arguing, advocating, and challenging—to ensure a free and appropriate public education for their children. It is noteworthy, however, that there are many other parents who choose to disengage from the educational decision-making process, despite their right to participate. What might parents have to tell *us* about how *they* experience the particular way that public schools in America have chosen to respond to disability?

A View of Special Education from the Margins

In this four decade collective narrative, 15 mothers describe their experiences within the institution of special education. Mothers do not bear the label of disability (although some may have been so labeled during their school careers); yet, they often interact as intensely with special education services as do their children with disabilities. Given their positionality within this system, mothers intimately experience both intended and unintended consequences of special education practices.

I chose to interview mothers (rather than fathers or sets of parents) because I have noted over the years that mothers are far more likely than fathers to engage with school personnel and/or with greater frequency. Despite the greater presence of mothers in schools as compared with fathers, mothers observe that school personnel typically show more respect to fathers on the basis of their perceived male rationality and objectivity. Furthermore, mothers report that parenting a child labeled LD inevitably becomes intertwined with expectations for "the ideal mother" in America (Warner, 2005)—an imposed cultural identity that fathers do not share. Thus, it appears that gender is a factor with which mothers must contend in negotiating with school personnel.

During my many years as a teacher and an educational diagnostician/ consultant, I learned that parent and family advocacy is as much about being a good listener as it is about being an agent of change. I listened to countless stories told by mothers. I remain struck by the sense of relief mothers expressed in response to having an opportunity to tell their stories. It was not the narrative particularities that were striking to me, but rather the uncanny degree of similarity. Such commonality of experience points toward the presence of issues that extend beyond a particular child, parent, teacher, principal, psychologist, school, or district.

Let us also not ignore the role of parents within the origins of the special education system. Parents speaking out for their children who could not speak for themselves led the movement toward a free and appropriate education for *all* children. Their efforts resulted in the 1975 passage of PL 94–142. Not only were children with disabilities guaranteed an education, but their parents were guaranteed participation in educational decision making. How is it, then, that parents—whose voices integrally contributed to the creation and passage of the law—report feelings of invalidation and powerlessness within a system designed to ensure their participation?

It is not my intent to dismiss any positive outcomes of special education practice in the lives of families nor to diminish the successful experiences of collaboration that mothers report. Within this analysis, I do, however, focus upon more clearly understanding *how* special education discourse *can* operate in ways that produce unintended (and largely unrecognized) consequences for families. Specifically, these 15 mothers speak about how they perceive dispositions of power within special education discourse, the ways they have positioned themselves within and against special education discourse, and the consequences of special education discourse in their lives.

Significance of Narrative Telling

It has long been documented that narrative structures, such as myths, fairy tales, legends, and stories, exist in all human cultures (Bettleheim, 1976; Bruner, 1990; Campbell, 1949). As Egan (1991) explains, "We are a storying animal; we make sense of things commonly in story forms; ours is a largely story-shaped world" (pp. 96–97). Similarly, Polkinghorne (1988) asserts that narrative is *the* fundamental way in which human beings understand life as meaningful

> [and] is a scheme by means of which human beings give meaning to their experience of temporality and personal actions. Narrative meaning functions to give form to the understanding of a purpose to life and to join everyday actions and events into episodic units. It provides a framework for understanding the past events of one's life and for planning future actions. (p. 11)

The attractiveness of narrative, then, lies in its illumination of *how* human beings make sense of their lived experiences, thereby deepening our understanding of the complexities of everyday life and human action.

My decision to seek out narratives about how mothers experience special education practices reflects an underlying assumption that "the important reality is what people perceive it to be" (Kvale, 1996, p. 52). Eliciting stories from 15 mothers across and within diverse generations, social classes, and race/cultural positions enabled me to construct a collective narrative that acknowledges the "indefinite series of perspectival view... [for] what is perceived is always incomplete because it is seen from a certain vantage point and there is always something unseen" (Merleau-Ponty, 1964, p. 106). In other words, this collective narrative documents the ways in which these particular mothers make sense of special education from within their diverse perspectives.

I believe that human beings can be understood only in relationship to their interactions with culture and others. Thus, the intersections between the social locations of these mothers (gender, race/culture, class) and the historical and cultural time period in which their narratives take place become central to this project. I agree with the Russian linguist and philosopher Mikhail Bakhtin who contends that there is no self without the social and cultural milieu for "without the environment to engage and test its capacity to respond, it would have no living existence" (Clark & Holquist, 1984, p. 66). Thus, a person's sense making of the world is necessarily intertwined with her social, cultural, and historical location.

The Mothers

In this section, I describe my method of participant solicitation as well as the social, historical, and racial/cultural composition of the participant sample. Given my purpose of constructing a historical account of special education discourse from the viewpoint of 15 mothers with children labeled LD, it is relevant to highlight group features of the participant sample. However, I am acutely aware that a compilation of group characteristics obscures the rich and nuanced complexity of individual narratives as well as my own influence as a researcher upon this narrative work. Thus, I begin chapters three–five with participant snapshots that illustrate the *particularities* of each mother's story featured in the chapter. I also clarify the nature of my relationship with each mother as an acknowledgment of my participation in the co-construction of these narratives.

The stories in this book emerged from individual interviews with the 15 mothers. Five of them also volunteered to participate in a subsequent group interview. I did not seek out mothers on the basis of any particular configuration of negative and positive experiences they may have had with school personnel nor did I solicit mothers with strong backgrounds in advocacy. I selected mothers on the basis of their interest in and availability for participation.

The participants reside in a total of four states, two in the northeastern United States and two in the southeastern United States. Geographical diversity exists among the participants, who come from rural America, small towns, medium-sized cities, and a large metropolitan area (see appendix A). I sought a degree of geographical diversity among participants for two reasons. First, I am interested in the experiences of mothers who are located in diverse positions of race, class, and culture. Having lived in both the southeastern and northeastern regions of the United States, I can attest to cultural variations between the two regions that exist beyond family heritage. Second, the Individuals with Disabilities Education Act (IDEA), a federal law, requires each state to submit a plan for IDEA implementation. Every state, then, must provide evidence of compliance with the federal law; however, bureaucratic implementation of the law may vary somewhat from state to state.

Of the 15 participants, 12 live in the southeastern United States and 11 of the 12 live in the same state. Since I spent 20 years working in the southeastern United States, I reached out to mothers, for inclusion in the study, with whom I had had a previous professional relationship—mothers I knew through my public school teaching, mothers whose children attended a private school for children labeled LD that I cofounded and directed, and mothers who sought services at the developmental pediatrics center where I worked as an educational diagnostician/consultant.

Of the 12 mothers who live in the Southeast, there are two participants with whom I did not have a previous professional relationship. One participant (who is also my relative by marriage) expressed interest about participating in the study as she has consulted informally with me about her child since his preschool years. The other participant, who lives in another state in the Southeast, was referred to me by a relative who had knowledge of my study.

I interviewed two mothers who live in a large metropolitan area in the northeastern United States. One mother had participated in a previous research project with me and was eager to be a part of this study. The other participant was referred to me by a former graduate student of mine who teaches in the public school system. Lastly, I interviewed a mother who lives in a small city in another state in the northeastern United States. She was introduced to me by a mutual colleague.

As stated earlier, I sought out mothers of diverse generation, race/culture, and social class to participate in the study. In the next section, I outline the configuration of these factors among the participants.

Generations. In that services for students labeled LD were not mandated within public schools until the mid-1970s, the smallest subset of mothers in the study are those whose children were born in the 1960s and 1970s. Reflective of the rise in identification of children as LD following the implementation of special education law, two-thirds of the mothers in this study have children who were born between 1980 and 1990—five with children born between 1980 and 1985 and five with children born between 1985 and 1990 (see appendix B). It is worth noting that I did not seek out a specific number of mothers to represent each generation, yet the availability of mothers reflects the upward trend of children identified as LD.

Race/Culture. Of the 15 participants, 13 are White (including one mother who self-identifies as Italian American and one mother who identifies her cultural background as Appalachian). One mother self-identifies as Hispanic, and one mother is African American (see appendix C).

Historical evidence indicates that students served within the learning disabilities category between 1963 and 1973 were mostly White and of middle-class status or higher (Sleeter, 1986). Thus, I anticipated that the participant pool would be predominantly White for those whose children were identified as LD during the years before the special education law was passed and the years shortly after the law was implemented. All participants from this era are indeed White and middle to upper class.

Given that I recruited participants with whom I had had a professional relationship during the 1980s and 1990s, it is noteworthy that all of these mothers also are White and identify themselves within a socioeconomic range of middle to upper class. These mothers represent the typical profile of parents with the economic means to enroll their children in a private school for children with learning disabilities and/or to seek psycho-educational testing and consultation at a private clinic. Moreover, the majority of these participants reside within an area of the southeastern United States whose population might be considered mildly diverse as compared with other areas of the country.

I had hoped to increase the racial/cultural diversity in this study by recruiting participants from the metropolitan area where I currently reside. I explored a number of avenues for possible referrals, including contact with colleagues, my current and former graduate students who teach in the public schools, and personnel within the public schools with whom I have worked. Having had years of experience in working closely with mothers of children labeled LD, I understood the significance of trust as a factor in whether or not mothers would choose to reveal their stories to me—a person who identifies as a special education professional. In the end, I was able to arrange and conduct an interview with an African American mother and a mother who self-identifies as Hispanic. I appreciate the contributions of their voices to my study, but regret that each is a sole voice. I understand that each mother speaks from her personal experience (as do *all* of the mothers in the study) and does not represent the experiences of all African American or Hispanic mothers.

Social Class. The 15 participants represent a range of social class. I asked each participant to place herself within one of the following categories: upper class, upper-middle class, middle class, lower-middle class, or at or below poverty level (see appendix C).

I did not assign a specific income range for each category because I was interested in how the participants generally perceive themselves in regard to social class.

Two mothers identify themselves as upper class. The largest group fell within the upper-middle class and the middle class, with four participants placing themselves in the former category and six in the latter category. One mother identifies herself as lower-middle class. The two remaining mothers place themselves at or below poverty level.

Mothers: A Valid Source of Knowledge

My method includes both individual interviews and a group interview in which mothers, in the course of their dialogue with one another, identify points of intersectionality among their narratives. Although I am able to construct a collective narrative from the individual interviews, I do so from the perspective of a researcher struggling to make sense of data. Data from the group interview, however, provide insight into mothers talking across time, context, and generation with one another about their respective experiences.

Individual Interviews. Of the 15 mothers interviewed, 12 requested to be interviewed in their homes. The remaining three mothers chose alternative spaces, including a school counselor's office, a spouse's empty office on a weekend, and a personal office space after work hours. I asked participants to commit to a 60–90 minute audiotaped interview. The average length of the interviews was between an hour and a half and two hours, with the shortest interview lasting 45 minutes and the longest lasting four hours. I did not to limit mothers to a specific length of time to tell their stories, but rather opened a space within which they could express themselves at a comfortable pace.

I opened each interview by asking the participant to talk about her child's learning disability as well as her experiences with special education. I also prepared a list of prompts to use during the interviews as necessary. It is noteworthy that nearly every mother addressed the following questions within her narrative without any prompting:

- How would you describe the language used in school meetings?
- How do school personnel speak about your child and his or her learning disability?
- How do school personnel speak about learning disabilities, in general?
- In your experience, who has the power in special education committee meetings and/or who makes decisions regarding the education of your child?
- What particular experiences support your perception of who holds the power in special education committee meetings?
- In what ways has your own knowledge of your child been received by school personnel?
- In what ways, if any, do you see your race/culture, social class, and/or gender impacting your relationship with school personnel?
- How has parenting a child labeled LD impacted how you see yourself as a mother?
- In what ways might this experience differ from the experience of raising a child who is not labeled?
- What do you see as the consequences of special education discourse for yourself, your family, and your child?
- What messages have you received from society about raising a child labeled LD?
- How might your lived experiences inform special education policy and practice?

Group Interview. The five mothers who participated in the group interview represent a subset of the 15 mothers who were interviewed individually. Each of the participants lives within a particular geographic region of a southeastern state. I recruited the members of this group on the basis of their availability and willingness to participate as well as their diverse social and generational locations (see appendix D). I intentionally arranged for the group to be small for the purpose of generating in-depth conversation among the participants.

One participant volunteered her home (centrally located in relation to where the other participants live) as the setting for the group interview. I began with an introduction of my relationship with each participant, followed by a review of the purpose of the group interview. As a way to initiate and circulate conversation among the participants, I asked each mother to draw a sketch that would represent how she envisions her past and/or present relationship with school professionals regarding her child's education. I distributed drawing paper, pens, and pencils to the group. After the participants completed their drawings, I asked for a volunteer to share her drawing with the others. In turn, each mother explained the meaning of her drawing and, in doing so, spontaneously told narratives of particular significance to her. As each mother took the floor, the others listened intently and actively, asking questions and engaging themselves in her story.

Following the drawing exercise, I asked the group to dialogue with one another about their visions for improving both general and special education as well as ideas for fostering authentic collaborative relationships between parents and school professionals. Lastly, the participants explored possibilities for disseminating their suggestions in ways that could impact special education policy and practice.

I had asked the participants to commit to a two-hour group interview; however, the interview lasted more than three hours. It is noteworthy that the participants chose to continue the interview because they decided there was more to discuss with one another.

Understanding and Portraying Narrative Meaning

My approach integrates narrative inquiry with critical discourse analysis (CDA). Through individual interviews, I elicited narratives from mothers about their experiences within the institution of special education. Drawing from this collection of narratives, I was able to map the experiences described by mothers onto a critical history of special education. I analyzed selected narratives using CDA, a method particularly suited to revealing relations of power, as a means to illustrate the ways in which mothers are positioned, yet not necessarily determined, by special education discourse.

Narrative Analysis. Following each individual interview, I transcribed the audiotape of the session. In each of the transcriptions, I identified both verbal and nonverbal nuances in an effort to better capture the essence of each participant's narrative. For example, I notated pauses, word emphasis (represented in italics), laughter, voice tempo and intensity, and body language (represented in bold print). The following excerpts from two transcripts illustrate my notation system.

Excerpt #1:

He did tenth and eleventh grade in that program in the public high school. And did so well that for twelfth grade, for the *first* time in his life [**voice breaks**], he went into *mainstream* [**voice full of emotion**]! I'm sorry [**slight pause, looks away**]. We *never* thought he'd been mainstreamed, but he did get *one* year of mainstream for twelfth grade.

Excerpt #2:

But they had tested him sitting in front of this window which opened onto a playground [**ironic laughter**]! This is where he is supposed to get his IQ and I *knew* that, you know, there were some *attention* difficulties [**laugh**], too. *But* I was very surprised when they told me that he had an IQ of—I believe it was—91. And they said that, you know, he was just the *sweetest* little boy, had the nicest blue eyes, but that he was just really *not that smart*. . . And he would be starting school at almost seven and a half. So [**sigh**], you know, I tried to balance between accepting my child for who he was and not really *believing* that this could be *right* [**tone of incredulity**].

Using Labov and Weletzky's (1967) definition of narrative, I identified and marked passages within each transcript that I considered to meet the authors' criteria for classification as narrative. Labov and Weletzky contend that five components signal the presence of narrative within a speaker's verbal expression: (1) Abstract (a brief summary of the story); (2) Orientation (time, place, persons); (3) Complicating Action; (4) Evaluation, Result, or Resolution; and (5) An Optional Coda (which returns the speakers to the present).

The following is an example of a transcript excerpt that I identified as a narrative according to Labov and Weletzky's criteria:

Abstract: It was a *real* miracle.

Orientation: Time—Pete's infancy
Place—doctor's office and home
Persons—doctor, Mimi, Pete, Mimi's mother, Hattie

And so, I went to the doctor and I said, "He's losing ground." Of course, the doctors were checking on him in Greenburg and in Pleasantville; they were checking on him constantly. I said, "He is losing ground and I can't keep the food down and I don't know what I'm going to do. We're not getting anywhere." The doctor said to me, "Mimi, you have got to let your mother have Pete and let her keep him for a while." Mother had offered several times. I said, "I can't. Nobody can feed him but me! I'm the *only* one. He'd be dead if I hadn't been feeding him."

Complicating Action:

And he says, "Well, I'm just telling you, *you* look to me like *you* really can't go much longer at this rate. You've got so much going on and this child is not going to *die.* Give him to your mother for about a week." So, I called my mother when I got home. I was crying. I was really upset and *extremely* tired . . . I called my mother and my mother said, "Mimi, doesn't Pete have a cold?" I said, "Yes." And she says, "Well, I can't take him. I'm *so* afraid that something will happen to him while he is at my house. I cannot take him [**voice lowers**]." And I became very *angry* with her! I didn't *tell* her I was angry, but I hung up the phone and I cried *my eyes out* [**with emphasis**]. I just cried and cried. And I thought, "What am I going to do?"

Evaluation, Result or Resolution:

. . . And Hattie, my maid, and I had gotten to where we would put him in the highchair and put pillows behind him because he couldn't sit alone. He was *still* not able to sit alone, turn over, or anything. But we would prop him in the highchair . . . He would eat, open his mouth . . . But *always*, there would be this [**makes gagging sound**] and it would all come back. So, I sat down to feed him . . . And I was at the end of my rope. And I *reached* over and I can remember now *grabbing* him *forcefully* and *jerking* him up out of the highchair and *plopping* him in my lap and *screaming*, "You've *got* to eat or you're going to *die*!!" You could have heard me, I am sure, out at the street [**voice lowers**]. The maid came running, "What is wrong?! What is wrong?!" I said, "He's going to *die*! He's *got* to eat!!" And I began to sort of *lose* it, you know? So, I got myself straightened out, propped him on my lap, reached over, and started to feed him the food I had been working on before and *he ate it.* And he did not throw it up. And he never threw up *again.* That was *it.* It happened *right* then . . .

Coda:

. . . That's *exactly* what happened. I *screamed.* I *screamed* at him. But *anyway*, we had an appointment to go back to [the hospital] that was about two months from then.

Once I identified a narrative within a transcription, I chose a title for the narrative that reflected its essence. For example, I named the preceding narrative The Miracle. During the data analysis phase, the naming of narratives proved useful to me in two ways. First, the act of choosing a title helped me to distill the core meaning of each narrative. Second, in culling and sorting through my voluminous data set, I relied upon the titles as reminders of narrative content while cutting and pasting the narratives into categorical groupings.

In studying each transcript, I paid close attention to the historical, cultural, and social contexts in which the narratives are set. I also considered the episodes recounted in each narrative within and among other stories told by the participant. Finally, I identified points of connection as well as points of tension across the narratives told by participants of diverse generation, class, and race/culture.

Given my interest in the historical/cultural/social positioning of the mothers, I identified and named three historical time frames within which to organize the narratives: (1) Special Education: The Early Years (1960s to mid-1980s); (2) Special Education: The Implementation Years (mid-1980s to mid-1990s); and (3) Special Education: The Maintenance Years (mid-1990s to 2004). I chose my own demarcation of historical time (rather than assigning narratives by traditional historical decades) to locate narratives within the breadth of historical moments significant to special education's evolution within public schools. For example, Congress passed PL 94–142 in the mid-1970s, requiring states to implement the law by the late 1970s—the immediate effects of which reverberated into the early 1980s; thus, I designated a time frame to accommodate the serial narratives spanning this particular time period of significance in special education in contrast to merely ordering narratives according to decade. Moreover, I used these historical demarcations to group the mothers by generation to compare and contrast their experiences over time. It was not feasible to group mothers by historical decade because children attended school during at least two decades and sometimes three (e.g., a child might start school at the end of one decade, attend through the entire next decade, and graduate at the beginning of the third decade). Thus, I placed mothers into generational groups according to the decades in which their children were born rather than by the decades during which their children attended school, resulting in an even distribution of participants (five mothers per generational group) in each historical time frame. This even distribution of participants made manageable the task of comparing and contrasting life experiences within generational groups.

Of the multiple ways I considered grouping the narratives, my chosen method represents the most workable solution (see appendix E). The initial time frame, Special Education: The Early Years (1960s to mid-1980s), represents the era before and immediately after the passage of PL 94–142 and includes narratives of mothers whose children were born in the 1960s and 1970s and attended school during the mid-1960s to the mid-1990s. My decision to group mothers of children born across two decades resulted in a longer time period during which the children could have attended school. For example, the oldest child in this group (who repeated a grade) began school in the mid-1960s and graduated in the late 1970s, while the two youngest children in this group began school in 1982 and graduated in the mid-1990s. I address this discrepancy by placing narratives pertaining to events through the mid-1980s in the initial time frame, Special Education: The Early Years (1960s to mid-1980s), and placing narratives pertaining to events beyond the mid-1980s in the middle time frame, Special Education: The Implementation Years (mid-1980s to mid-1990s). Thus, narratives of the last two mothers introduced in the initial time frame appear in both the initial and middle time frames, depending upon the historical context in which the narrated events occurred.

The middle time frame, Special Education: The Implementation Years (mid-1980s to mid-1990s), corresponds with the years after the passage of PL 94–142 in which states focused upon the law's implementation, an era in which both schools and parents tested the limits of the new law. In this time frame, I introduce mothers whose children were born between 1980 and 1985 and attended school during the mid-1980s to 2002. The narratives collected from this group of mothers primarily take place between the mid-1980s and mid-1990s; however, a few narratives occur beyond the mid-1990s as this group of children graduated from high school between the latter part of the 1990s and 2002. Thus, narratives set beyond the mid-1990s and told by mothers in this generational group appear in the subsequent time frame.

The final time frame, Special Education: The Maintenance Years (mid-1990s to 2004), refers to the most current era in which special education, a public school institution now for nearly 30 years, is the naturalized educational response to students who demonstrate difficulty learning. In this last time frame, I place narratives of mothers whose children were born between 1988 and 1990 (no children whose mothers participated in the study were born in 1986 or 1987) and attended school from the mid-1990s through 2004. In addition, two mothers introduced in Special Education: The Maintenance Years (mid-1980s to mid-1990s) also appear in this historical time frame because each has a younger child who, like her older sibling, receives special education services.

Critical Discourse Analysis. Out of this collection of interviews, I selected narrative samples from each historical time frame to analyze using CDA. Fairclough (1989) defines CDA as an analytic tool that increases

> consciousness of language and power, and particularly how language contributes to the domination of some people by others... this means helping people to see the extent to which their language does rest upon common-sense assumptions, and the ways in which these common-sense assumptions can be ideologically shaped by relations of power. (p. 4)

Thus, CDA is an analytic tool particularly well suited for my purpose of exploring how mothers become positioned, yet not necessarily determined, by special education discourse.

According to Fairclough (1989), if we conceive of language as a form of social practice, we more clearly see "conventions routinely drawn upon in discourse [that] embody ideological assumptions which come to be taken as mere 'common sense'" (p. 77). An analysis of discourse, then, necessarily includes the examination of *subject* positions within that discourse, an aspect particularly relevant to the study of mothers and their relationship with special education professionals. Thus, it is within the discourse of special education that I looked for evidence of *how* mothers of children labeled LD become positioned. In my efforts to better understand this phenomenon, I combed the transcripts for examples of particularly rich narratives that demonstrated (1) how mothers of children labeled LD report being positioned within or against the discursive practices of special education; (2) what such positioning reveals about the ideology of learning disabilities as practiced in special education; (3) how subject positioning influences the way mothers with children labeled LD view themselves as mothers; and (4) how mothers come to resist subject positioning. Thus, the narratives chosen for CDA reflect one or more of these four criteria.

In applying CDA to the selected narratives, I first clarified the situational context of each narrative by identifying what is going on, who is involved, what relationships are at issue, and what role language plays in the episode described (Fairclough, 1989, p. 147). Subsequently, I read and analyzed each narrative at three levels of text. First, I focused upon the word level of text, noting classification schemes, use of ideologically contested words, rewording or overwording, euphemisms, metaphors, markedly formal or informal words, and expressive values of words. Then, I reread each text at the grammar level, looking for examples of agency within sentences; nominalizations; positive or negative wording of sentences; modes of grammar (declarative, grammatical question, imperative); use of "we" or "you"; and ideas contained within subordinate clauses. Lastly, I read each text for indications of interactional conventions and ideological assumptions held within larger society, paying particular attention to the historical time frame in which the narrative takes place.

As to be expected in a conversational setting, overlaps in speech occurred during the group interview. Such overlaps are of interest as points of validation as well as tension among the mothers. When overlapping dialogue occurred, I notated this by transcribing the overlap below the word(s) in the previous speech utterance in which the overlap occurred, as in the following example:

Katie: And
10 here's the cheering section. This is the Normal Child and this
is the LD Child. There's *all* kinds of people cheering that normal

child on. Helping them, supporting them, guiding them. And that poor little LD Child.

Kim: And cheering the *mom* on, too [**group laughter**]!

15 Katie: Yeah, cheering the mom on, too, for doing such a good job! And there's the LD Child. And he just has piddly two people over there. That's how I feel. I feel like there's *so* much for the Normal Child and there's just so

20 few people out there that grasp what that LD Child needs.

Lindsey: And they don't even *see* all these hurdles.

Katie: They don't! And they don't really *care*!!

25 Lindsey: Uh-uh!!

Katie: That's the hard part. The people that don't *care* [**voice lowers**].

A Caveat or Two

In light of my tremendous investment in this problem and extensive background in working with mothers, I acknowledge that I came to this study with ideas about the kinds of narratives I would hear. Moreover, I recruited a number of mothers for the study with whom I had had a previous professional relationship. Thus, I was familiar with some of the stories that mothers told and/or I had been present (as an advocate) at some of the special education committee meetings described. Although this might be considered a liability in a traditional research design, I counter that my *lack* of objectivity in regard to this problem as well as my close relationship with most of the participants contributed both depth and complexity to this qualitative study. For example, it is worth noting that the narratives of mothers with whom I have an established history are noticeably richer and more deeply nuanced than the narratives of the two mothers I met for the first time at the interview.

To construct a historical account of special education discourse from the viewpoint of mothers, I asked participants to rely on their memories of past events, some of which may have happened over 40 years ago. Although questions could be raised regarding the reliability of any person's memory for past events, I am, as a qualitative researcher, less concerned with historical accuracy and more interested in the *perceptions* mothers hold in regard to their recalled experiences. I consider this qualitative study, like *all* texts, to be partial and situated. I do not make claims to grand generalizations or to telling "the whole story." I offer instead what I believe is an evocative work that adds to our knowledge base. I concede that stories *about* what happened are not identical to what *actually* happened in a particular time and place. I argue, however, that the events mothers recall are valuable representations of their experiences. Furthermore, I consistently observed throughout the interview process that mothers recall significant events in regard to their children with both remarkable detail and emotional intensity.

This study is limited in the sense that I gathered data from a representative group of mothers. Such a limitation, of course, is endemic to nearly all research studies. Of greater concern is the minimal representation of racial/cultural diversity within the study. As previously discussed, I regret that I was not as successful as I had hoped to be in recruiting mothers of color. I acknowledge that this book primarily reflects the experiences of White mothers.

Finally, as I expose, validate, and challenge the uneven power differential between professionals and mothers, I acknowledge my own participation in this power structure because of my positionality as a special education researcher writing about the narratives told by mothers. As sincere as I believe my intentions to be in representing these mothers as they wish to be represented, this collective narrative, nonetheless, is an account filtered through my eyes as a special education professional and retold through my words as a researcher. Nonetheless, it is my hope that this book will contribute to a deeper understanding of the mothers who navigate our system of special education.

Chapter Two

From Past to Present: American Culture, Public Schools, and Parents

Given my project of constructing a historical account of special education discourse from the viewpoint of 15 mothers with children labeled learning disabled (LD), it is relevant to consider this history within the context of the documented history of public schools, special education, and the field of learning disabilities. It is not enough merely to tell the story of these mothers. Their lived histories make sense only within the framework of the professional discourse that positions them—a discourse that provides a context for *why* mothers react and *what* they resist. Thus, I examine knowledge deemed "official" for two reasons. First, I believe such a historical review provides the backdrop to more fully understand the positioning of mothers within special education discourse. Second, it is against this historical review that I overlay the mothers' narratives.

I begin by looking at the evolution of public school education in the United States and the subsequent ways in which parents, particularly mothers, have become positioned within the discourse of public schools in general and over time. Given that special education exists *within* the institution of public school, it is relevant to explore the historical relationship between parents and public schools to more fully understand the response of public school professionals to the unprecedented parental rights afforded by the Education for All Handicapped Children Act (PL 94–142). In mapping a critical history of the relationship between parents and public schools, I address the following questions: How has public education discourse shifted over time? How have these shifts influenced the relationship between parents and public schools? What cultural and societal influences are reflected in these shifting truths? What circumstances fostered the acceptance and distribution of particular discourses? How have parents responded to particular discourses through time?

In looking at the history of public schooling in the United States, it is significant to consider what factors led to the construction of special education—and more specifically the field of learning disabilities—at a particular time within a particular social and cultural context. How did the shifting discourses of public education make it possible for the learning disabilities field to emerge in its current form as an educational alternative for parents and their children? I present two perspectives about the emergence of the field—learning disabilities as a scientific discourse and learning disabilities as a social and political discourse.

As an institution, special education revolves around the classification of children on the basis of their "special needs." Given the assumption that children could and should be sorted according to ability, there is also the implicit assumption that educators have at their disposal a reliable means by which to conduct this sorting. This assumption takes on particular significance in regard to the category of learning disabilities, which is, by nature, more elusive and subtle than some other disability categories. It is, therefore, significant to consider the cultural construction of normal/abnormal as well as the rise of testing, particularly as it relates to the apparatus of special education. What is the historical basis for our assumption that some children *naturally* fall outside the range of what is considered "normal?" How might these assumptions drive the discourse of special education?

There is a slender body of literature about parent and professional collaboration under the Individuals with Disabilities Education Act (IDEA) guidelines. This literature primarily offers information *about* mothers (rather than reports *from* mothers) regarding their collaborative experiences with professionals. I address the literature in two sections—the years shortly after the implementation of PL 94–142 (1978–1985) and the subsequent years (1986 to the present).

Finally, I analyze the discursive practices of special education. In my effort to understand more clearly the relationship between mothers and school professionals within the context of IDEA, I consider three theoretical lenses through which to explore the discursive practices of special education—power and parent status, power and parent gender, and power and race/class.

A Critical History of the Ongoing Relationship between Parents and Public Schools

In this exploration of the ongoing relationship between parents and public schools, I choose a critical stance as a means to examine the professional discourse that has enveloped parents throughout the evolution of public schools. Given my intent to construct a historical account from the viewpoint of mothers of children labeled LD, a critical history of American public schools serves as the backdrop against which to consider these narratives. Moreover, I believe that a critical history might begin to center the historical margins by fleshing out the official rendering of how the field of learning disabilities emerged and developed.

This chapter begins with an abbreviated history of the evolution of American public schools. Within this history, I identify institutional structures, social and political events, and economic practices and processes that opened space for certain discourses to exist and not others. In particular, I explore the ways that parents, particularly certain groups of parents, have become positioned and repositioned within the discourses of public schooling. This brief historical overview spans the early to mid-twentieth century and concludes with a more in-depth focus upon the 40 years relevant to this study—the 1960s to the present.

Parents and Public Schools: Mid-Nineteenth Century to Mid-Twentieth Century

The roots of public schooling can be traced to the establishment of democracy as the defining governing structure of the United States. Early lawmakers, acting upon the national discourse of personal freedom, responsibility, and empowerment, eventually establish a system of free and public education for the nation's children. As we shall see, it is the democratic notion of "rugged individualism" (i.e., pulling oneself up by the bootstraps)—perhaps the most defining feature of the American ethos—that repeatedly plays out in public schools in ways that hold parents and their children responsible for their own success or failure (Varenne & McDermott, 1998).

"The melting pot of America," a textbook phrase well known to generations of schoolchildren, portrays the image of a democratic and embracing America offering her land of opportunity to all people. However, to lift the lid of the melting pot is to expose a brewing, boiling, sputtering stew of American society at the turn of the twentieth century in which members of a dominant culture (White, Anglo-Saxon, Protestant) make strategic responses to preserve their privileged status within an increasingly diverse society (Anderson, 1988; Tyack, 1974). Public schooling, then, becomes "part and parcel of a national morality play in which those hopes and fears were enacted" (Kliebard, 1995, p. 251).

What specific historical contingencies account for public schooling emerging as a primary site from which to disseminate discourses of the dominant culture? At the turn of the twentieth century, an increasingly complex social, political, and economic landscape defines America. Population shifts occur as industrialization brings people from the country into the cities (Kliebard, 1995). The social and economic needs of a burgeoning immigrant population challenge America's cities (Rousmaniere, 1997). Science penetrates American society, giving rise to scientific management of factories, a new class of scientific professionals, and scientific study of human beings (Gould, 1996; Kliebard, 1995). American nationalism rises in the wake of World War I, resulting in heightened suspicion and distrust of immigrant populations (Tyack, 1974). Moreover, low-status groups, such as African Americans and Native Americans, face a hostile society that controls access to educational, cultural, and economic capital (Anderson, 1988; Foster, 1997; Lomawaima, 1995).

Given the complexity of these political and social issues, dominant culture leaders (White, Anglo-Saxon, Protestant) view school reform as a viable and effective avenue for preserving their privileged status in American society and managing the influx of an increasingly diverse population. It should be noted that the progressive educational reform of this historical period is best understood as "shifting coalitions around different issues" (Kliebard, 1995, p. 240); however, historians and academicians generally agree that four major academic discourses (humanism, developmentalism, social efficiency, social meliorism) emerge by the turn of the twentieth century. Although all four discourses contribute to the evolution of public schooling, social efficiency becomes the discourse with the most long-ranging and consistent impact upon American schools (Kliebard, 1995), particularly in regard to special education—a point I develop further in a subsequent discussion of the rise of testing.

Where might parents, particularly mothers, be positioned within the early discourses of public schooling? According to Cutler (2000), the advent of bureaucratized schools

> drove a wedge between the home and the school that would widen as the nineteenth century progressed. Increasingly, parents and teachers faced each other across a gap that featured systematic procedures and standardized expectations. Report cards replaced more personal forms of communication. Graded schools became the norm...the development of teachers' institutes and normal schools gave rise to the belief that esoteric knowledge made teachers uniquely qualified to oversee the education of the young. (p. 17)

Thus, beginning as early as the 1840s, the discourse of schooling gives rise to the *institution* of education. The newly organized apparatus of American public education establishes structures around *who* has the right to make statements about education, *what sites* these statements originate from, and *what positions* the subjects of discourse inhabit (Dreyfus & Rabinow, 1983). Educators assume a position of superiority on the basis of their professional training, allowing them to claim greater and more scientific knowledge than parents. It is noteworthy that, at the outset, the institution of education firmly establishes systematic procedures and standardized expectations, signifying that input from parents would be neither sought nor valued (Cutler, 2000).

Throughout the early to mid-twentieth century, educators continue to define the position of parents within the discourse of schooling. For example, mothers participate in school-sanctioned groups, such as the National Congress of Mothers, Parent Teacher Associations, and other like organizations, whose primary functions are "performing menial tasks, supplying school libraries, or improving facilities" (Valle & Reid, 2001, p. 25). In his study of home/school relationships of the Progressive Era, Reese (1978) asserts that administrators and teachers consider "home-school cooperation [to be] a fine thing in theory, it seemed, as long as parents behaved properly... even an indifferent parent was preferable to one who repeatedly clashed" (p. 20).

There is no mistaking the paternalistic positioning of school personnel in relationship to parents. Cutler (2000) characterizes the early to mid-twentieth century as a time in which the institution of school emerges as "an educator of parents as well as children" (p. 163). However, school personnel direct paternalistic attitudes even more intensely toward certain groups of parents. As schools increasingly take responsibility for assimilating immigrant populations into American society, social programs develop within schools to provide poor immigrant families with needed clothing, medical screening, and school lunches (Kliebard, 1995). Although such charitable actions no doubt improve the lives of poor and immigrant families, school charity also carries a clear message—to participate in American schools is to be assimilated into Anglo-Protestant middle-class culture and values. For the most part, school personnel as well as Anglo-Protestant mothers believe that "poor, Catholic, and/or immigrant parents are incapable of raising their children properly" (Valle & Reid, 2001, p. 27). It should be noted that school charity programs do not extend to African American families, despite the dominant culture's widely accepted perception of poor parenting skills and low-achieving children within this population (Anderson, 1988). Thus, the *degree* of school paternalism parents experience appears related to race, ethnicity, and social class—a pattern we shall see repeated throughout the history of home/school relationships.

As a result of America's involvement in World War II near mid-century, public schools take on an intensified role in "helping to create and maintain a democratic moral" (Smith, 1942, p. 113), shifting attention from the family to the nation. The life adjustment movement, rooted in the tenets of social efficiency, emerges in response to the needs of a wartime population. Thus, the American curriculum takes a decidedly practical turn (Kliebard, 1995). In the spirit of a nation pulling together under the threat of communism, school administrators "remind parents that they should be advocates for public education" (Cutler, 2000, p. 167).

In the postwar era, the "all for one" mentality of the 1940s gives way to an unprecedented criticism of public schools in the 1950s (Kliebard, 1995). Parents increasingly blame progressive education for undermining parental authority and decreasing academic standards, especially in light of the rising rates of juvenile delinquency (Cutler, 2000). Moreover, public education receives heightened criticism following the Soviet Union's successful 1957 launch of Sputnik, ultimately leading to curriculum changes and an emphasis on educating gifted and talented students (Rickover, 1959). In this era of Cold War, science moves to center stage as our nation's Discourse of Truth.

Parents and Public Schools: 1960s to the Present

The turbulent politics of the 1950s and 1960s shape American education in new and significant ways. In response to the civil rights campaigns of the 1950s and 1960s, the Johnson administration implements the War on Poverty, a federal program designed to disrupt intergenerational poverty. Under this program, the federal government establishes the Head Start and the Elementary and Secondary Education Act (ESEA) in an effort to improve educational outcomes for low-income students (Spring, 1989).

With the civil rights movement at the forefront of national consciousness, a resurgence of interest in democracy and equality predictably emerges within our national discourse. The federal government, responding to the tenor of such discourse, "endorses parent involvement in educational-policy making, requiring parent advisory councils for Head Start Centers funded under the Economic Opportunity Act" (Cutler, 2000, p. 176). In an effort to facilitate home/school relationships, mandatory home visits are required of all Head Start programs (Anselmo, 1977). Moreover, Head Start personnel actively encourage parents to observe and/or volunteer in the classroom as well as participate in weekly parent meetings about topics related to proper family life (Feldman, 1989).

Under the auspices of "parent involvement," programs such as Head Start *really* function to assimilate low-income and culturally diverse mothers into the dominant culture's idea of schooling and parenting (Spring, 1989). In keeping with historical precedent, the cause of poverty is believed to originate within the disadvantaged lives of children rather than within the unequal social and economic systems of the United States (Brantlinger, 2003). In other words, the discourse on poverty includes the assumption that low-income children would be better prepared to enter school if their mothers were taught to parent like White middle-class mothers.

Thus, the War on Poverty, with its emphasis on changing *parents* and not schools, effectively replicates the paternalistic school rhetoric evident in the first half of the twentieth century (Valle & Reid, 2001). Of further significance is the creation of a government technology through which to channel federal funds for the purpose of identifying and serving a "special" Head Start population (meaning deficient as compared with the rest of the school population) that needs "special" Head Start programs (meaning a different kind of education). With a precedent established for federally funded educational programs that target an identified school population, the groundwork is laid for a subsequent discourse to materialize around "special" education for students with disabilities.

Civil rights discourse opens space for African American and Latino parents in New York and New Jersey to lobby successfully for school decentralization and autonomy of school affairs. However, many school administrators resist parental input regarding curriculum, budget, and personnel and community school boards intentionally limit information shared with parents (Cutler, 2000). Bell (1978) attributes disappointing educational outcomes for African American students during the 1970s to

> the absence of black parents in making policy, the exclusion of black parents from their children's education, and the inability of black parents to hold school personnel accountable for meeting the educational needs of their children. (p. 117)

Although the civil rights movement opens space for parents of color to move into the discourse on schooling, school personnel maintain power by controlling the *nature* of parent participation.

The discourse of cultural deprivation, initiated at the 1964 National Conference on Education and Cultural Deprivation, persists well into the 1970s. The culturally deprived, identified as "Puerto Ricans, Mexicans, southern blacks and whites who moved to urban areas, and the poor already living in inner cities and rural areas" (Sleeter, 1987), become the target of scientific studies that document differences in mother/child interactions (meaning different from White middle-class mothers), thereby justifying intervention by educators who conduct home visits (Conant, 1971; Horn, 1970; Olmstead & Jester, 1972). By identifying so-called deprived home environments as the reason for school failure, educators circumvent responsibility for such failure (Valle & Reid, 2001).

During the 1970s, the federal government becomes increasingly involved in legislating parents' rights. For example, the Family Educational Rights and Privacy Act, passed in 1974, grants parents access to their child's school records and prohibits the release of school record information to a third party without parental consent (Cutler, 2000). The following year, Congress enacts the Education for All Handicapped Children Act (PL 94–142), representing the foremost example of parents' rights legislation to date. Building on the discourse of the civil rights movement and Brown versus Board of Education, parents of children with disabilities and their advocates lobby for a free and appropriate public education for *all* children. The law (renamed the Individuals with Disabilities Education Act or IDEA in 1990) not only guarantees a free and appropriate public education for children with disabilities, but also grants parents due process under the law. For example, parents have the right to be informed, the right to be knowledgeable about the actions to be taken, the right to participate, the right to challenge, and the right to appeal (IDEA, 1990, 1997, 2004). In accordance with the law, the institution of special education emerges as a public school requirement, a system within a system, to construct its own particular discursive practices.

Likewise emerging out of civil rights legislation and garnering widespread media attention throughout the 1970s, the women's movement introduces a discourse to challenge the second-class status of American women. Despite the significant impact of this discourse upon our national consciousness, public schools of the 1970s continue to hold traditional views of gender roles.

> PTA projects were often in keeping with assumptions about what women could do and liked to do. They included such such "female-oriented" projects as holding bake sales, organizing teacher appreciation days, and maintaining gardens at school. (Valle & Reid, 2001, p. 28)

Moreover, divorce rates climb throughout the 1970s as do the numbers of married and single mothers entering the work force; yet, school professionals persist in conceptualizing "family" as the traditional two-parent home with father working and mother at home, failing to acknowledge or address the needs of nontraditional families (McFall, 1974).

The discourse of the open classroom, popularized in American schools between the mid-1960s and mid-1970s, integrates rapidly expanding research on child development into classroom practice (Cuban, 1993). The concept of open classrooms primarily reflected the groundbreaking work of Swiss developmentalist Jean Piaget, who believed that "the learner *constructed* his or her own knowledge through the interplay of maturation, experience, and thinking about experience" (Reid, Hresko, & Swanson, 1996, p. 224, emphasis in original). In contrast to a traditional emphasis upon direct instruction and mechanistic learning, teachers in open classrooms facilitate instruction around active learning, interactions with others, and self-regulation.

The era of the open classroom, however, is short lived. By the mid-1970s, a Back to Basics discourse moves to center stage in the national arena. The source for this dramatic educational shift is somewhat unclear. Cuban (1993) suggests that such a discourse may have developed in response to "persistent reports of declining test scores, increasing school vandalism, disrespect for teachers, or the educational version of the newly conservative political climate" (p. 207). Furthermore, embedded within the Back to Basics discourse is the notion that *some* people's children, *unable* to benefit from the constructivist teaching in open classrooms, require, instead, highly structured instruction based upon the tenets of behaviorism (i.e., two types of children exist who need two distinct types of instruction). Thus, space opens for the eventual development of parallel systems of education—general education (influenced by child development and constructivism) and special education (influenced by behaviorism and technology).

Following a period of relatively high unemployment during the 1970s and early 1980s, the National Commission on Excellence in Education issues *A Nation at Risk: The Imperative for Educational Reform* (U.S. Department of Education, 1983), warning America that its global competitors (Japan, West Germany, South Korea) may overtake its position as a world leader in commerce, industry, science, and technological innovation. Thus, a discourse emerges that targets ineffective schools as the cause of the country's slow economic growth (Spring, 2002). Beginning in the early 1980s (and persisting through the present), this particular discourse becomes widely distributed through waves of state policies directed at increasing academic standards and enforcing minimum competency testing for graduation (Cuban, 1993). Varenne and McDermott (1998) contend that such enduring political focus upon ineffective schools leads to a naturalized practice in which

> bad schools and failing children are separated out. They are made to be "different." They are the target population. They are isolated to the point that the interaction between good and bad schools, successful and failing children, is lost... [there is] evidence for two worlds, two societies, separate and unequal, a dominant one to be emulated and a colonized other to be explained and transformed. (p. 109)

Thus, the discourse of standards and competency testing serves to reinforce the notion of *two* types of education for *two* types of children.

In keeping with the Excellence in Schools rhetoric of the 1980s and 1990s, reformers strongly encourage schools to adopt "site-based management" as an avenue for more effectively implementing school policy. In the context of site-based management, local school administrators share decision making with a committee of teachers, parents, students, and other community members. However, researchers document resentment on the part of some administrators who perceive such forced collaboration as an affront to their authority as well as a drain on their time (Evans & Perry, 1991), even though site-based management clearly "is not intended to give meaningful power to teachers, students, and community members [including parents]" (Spring, 2002, p. 162).

While the standards movement gains momentum in the 1980s and 1990s, multiculturalism emerges as a concurrent educational discourse in light of the rising influx of immigrants during this period (Nieto, 1999). From 1981 to 1990 alone, an estimated 8 million legal immigrants took up residence in the United States (U.S. Bureau of the Census, 1993). This estimate does not include, of course, the number of illegal immigrants residing within the country. Furthermore, diversity among these newcomers is significant, with immigrants representing countries in Latin America, the Caribbean, Asia, and Africa (Valle & Reid, 2001). Much like the nation's response to immigration in the early part of the twentieth century, members of America's dominant culture voice fears (via public and political debate as well as media circulation) regarding the negative consequences of immigration upon American culture and schooling (Valle & Reid, 2001). Given the long tradition of American schools as a vehicle through which to promote the majority culture's political, social, and economic agendas, students who bring diverse languages and cultures to school often struggle to learn, prompting school personnel to refer many such students for special education (Lagrander & Reid, 2000). Moreover, school personnel may hold mistaken assumptions about the abilities of immigrant students and their parents (Darder, 1991). Inspired and informed by the civil rights movement, proponents of multiculturalism define the goal of multicultural education as a means "to provide equality of opportunity in the existing economic system by reducing racial and cultural prejudice" (Spring, 2002, p. 41). In contrast to the notion of a standard curriculum, multicultural educators focus their efforts on creating curriculum to validate multiple ways of "being and knowing" in diverse cultures, promoting community in the classroom, and more effectively embracing immigrant children into America's schools (Nieto, 1999).

Throughout the 1980s to the present, feminist scholars, like their colleagues who promote multiculturalism, continue to identify ways in which our public schools consistently reinforce and elevate rational, patriarchal, and Eurocentric conceptions of knowledge above "other ways of knowing" (Grumet & Stone, 2000). For example, Belenky, Clinchy, Goldberger, and Tarule (1997) argue that the kind of knowledge that that is used in child rearing is

> typical of the kind of knowledge women value and schools do not... Good mothering requires adaptive responding to constantly changing phenomena; it is tuned to the concrete and particular... maternal thinking differs from scientific thinking. (p. 313)

Valle and Reid (2001) contend that "mothers, [in part] because they are women, are disempowered in the face of the objective, scientific, and bureaucratic institution of schooling" (p. 28). Furthermore, feminist scholars vehemently reject the notion that low-income mothers, particularly of nondominant cultures, are a legitimate source of blame for school failure—arguing that the root of school failure is instead located within "the social and economic structures in which mothers' lives are embedded" (Manicom, 1984, p. 79).

Despite the significant attention that proponents and scholars of multiculturalism and feminism attract within the educational community, research indicates that the actual impact within the monolithic bureaucracy of schools may be minimal. For example, critics argue that multiculturalism trivializes cultures when reduced to nothing more than a curriculum of decontextualized activities, such as "Brown Holidays and Heroes" (Banks, 1991; Nieto, 1995). Moreover, many educators continue to display blaming attitudes toward low-income and culturally diverse mothers who, in turn, opt to avoid interactions with authoritative school professionals (Kalyanpur & Harry, 1999). Heleen (1992) contends that a mutually reinforcing cycle exists between home and school because school personnel really do not expect low-income parents

> to participate productively in their child's educational life and [make] few efforts to build positive relationships; poor parents, in turn, [do] not expect to be meaningful actors in their children's learning and often [have] nothing but negative experiences with the school. (p. 6)

By the dawn of the twenty-first century, America's discourse on schooling is clearly dominated by liberal *and* conservative politicians alike who bandy about educational catchphrases such as "high standards," "school choice," and "charter schools"—seemingly in an effort to create a national discourse on schooling to which the public can subscribe (Traub, 2000). It appears that members of the majority culture intend to save our schools by imposing higher educational standards in the name of equalizing opportunity for all—a political stance particularly well reflected within the rhetoric of the No Child Left Behind Act (2001). As a nation, we might be wise to heed the caution of Paulo Freire, the influential Brazilian theorist, who states that "dominant elites...encourage passivity in the oppressed and take advantage of that passivity to fill that consciousness with slogans which create even more fear of freedom" (in Flinders & Thornton, 1997, p. 153).

Moreover, Apple (in Casey, 1993) argues that scathing criticisms of public schools can be seen as a way for members of the dominant social groups "to deflect a crisis often created by their own decisions onto schools and teachers" (p. xii), and I would add, onto families. If students fail to rise to the increased educational expectations held by teachers and schools (in other words, "choosing" not to help themselves), they will suffer the consequences that a lack of education brings. Not coincidentally, this discourse rings of the "personal responsibility" rhetoric associated with recent welfare reforms. Such a viewpoint neatly places blame for school failure upon students—blame that ultimately extends to their families—and absolves society of its contribution to the contextual realities of people living outside the dominant culture.

The Rise of Learning Disabilities within Public Schools

Having outlined a critical history of parent/school relationships within American public schools, I turn now to an examination of the social, historical, and cultural context that opens space for the learning disabilities field to materialize. In other words, in what ways did the shifting discourses of public education contribute to the emergence of the learning disabilities field as a viable explanation for children who displayed difficulty learning? I offer two discursive perspectives about the emergence of the learning disabilities field: learning disabilities as a scientific discourse (the "official" history of the field) and learning disabilities as a social and political discourse.

Learning Disabilities as a Scientific Discourse

It is widely accepted that the learning disabilities field was officially established in 1963 at a conference sponsored by the Fund for Perceptually Handicapped Children (Meyers & Hammill, 1990). Samuel Kirk, a prominent speaker at the conference, is credited with having introduced the term "learning disabilities" to differentiate a particular group of children with learning and social difficulties from other children with disabilities.

> I do exclude children who have sensory handicaps such as blindness, or deafness, because we have methods of managing and training the deaf and

> blind. I also exclude from this group children who have generalized mental retardation. (Kirk, 1963, p. 3)

Conference attendees voted to establish the Association for Children with Learning Disabilities, thereby creating an apparatus through which to introduce and distribute the burgeoning discourse of learning disabilities (Meyers & Hammill, 1990). To appreciate how this particular conception of a learning disability emerged, and not others, it is relevant to revisit the earliest attempts of scientists and medical professionals to understand the nature of reading difficulties in children.

By the turn of the twentieth century, scientists from diverse medical backgrounds begin to study children who display persistent problems learning to read. For example, James Hinshelwood, a Scottish ophthalmologist working in the early decades of the twentieth century, looks to medical studies of brain-injured adults and concludes that developmental reading problems result from congenital brain damage—a condition he terms "congenital word blindness" (Meyers & Hammill, 1990). By the 1930s, Samuel Orton, an American neurologist, challenges Hinshelwood's theory of congenital brain damage, asserting that developmental reading problems instead arise from an absence of dominance in the left hemisphere of the brain thought to regulate language. He names the condition "strephosymbolia" and, along with Anna Gillingham, develops an extensive instructional program to remediate its symptoms (Orton, 1937). Although Hinshelwood and Orton are recognized for their contributions to the field of learning disabilities, it should also be noted that subsequent research does not substantiate either theorist's work.

German scientists, Strauss and Werner, also working during the 1930s, focus their research upon children who display generalized behavioral and cognitive difficulties. Drawing upon behavioral observations of head-injured World War I veterans, Strauss and Werner presume brain damage to be the cause of similar behaviors in children (Hallahan & Cruickshank, 1973; Meyers & Hammill, 1990). Thus, the presumably brain-injured child with features of distractibility, disinhibition, perseveration, and dysfunction in visual perception and visual motor integration becomes classified as the Strauss Syndrome Child, a condition later renamed minimal brain dysfunction (MBD) (Meyers & Hammill, 1990; Sleeter, 1987). Working with Lehtinen in the 1940s and Kephart in the 1950s, Strauss develops instructional programs intended for brain-injured, perceptually handicapped children. Extending upon the work of Strauss and colleagues, Cruickshank applies similar techniques to instructing perceptually impaired, hyperactive children of normal intelligence (Hallahan & Cruickshank, 1973). Numerous researchers working in the 1950s and 1960s (e.g., Frostig, Getman, Barsch, Kephart, Delacato) develop perceptual-motor tests and instructional programs based upon Strauss and Werner's work (Meyers & Hammill, 1990).

Although subsequent research disproves Strauss and Werner's theory of perceptual-motor development as the basis for higher learning, it is important to consider the impact of this discourse upon the embryonic field of learning disabilities. It is noteworthy that the earliest studies of children with learning and behavioral difficulties come out of the medical community. As subjects of study for ophthalmologists, neurologists, scientists, and psychologists, children with learning difficulties become enveloped within a *discourse of science*. Professionals, operating out of a scientific framework, describe these children in terms of pathological deficits—a conceptualization that eventually will undergird the institution of special education.

Learning Disabilities as a Social and Political Discourse

I have, thus far, described an official history of how the learning disabilities field emerged, told as a continuous tale of scientific progress leading to the ultimate discovery of an identifiable and treatable childhood pathology. It is a history well documented in books, journals, and college texts—a history passed onto college and graduate students who seek to enter the field's professional ranks. If, however, we return to the social and political arena of the early 1960s (the moment in time when the learning disabilities field was "born"), we might consider *how* spaces opened to accommodate and distribute particular discursive statements. In other words, what social and political conditions allowed certain statements (presuppositions that circulate as true and accepted) to exist and not others?

Let us revisit the decade preceding the 1963 conference at which Kirk first used the term "learning disabilities." Postwar America stews with uneasiness over the threat of communism, a certainty to which our military, media, government, and most everyday citizens subscribe. Following the successful launch of Sputnik in 1957, the steadily mounting competition for control of worldwide military and business interests intensifies between the United States and the Soviet Union as does the idea that American schools must prioritize education for the academically gifted who will become our nation's scientific, business, and technological leaders (Cutler, 2000; Sleeter, 1987). For example, Rear Admiral H. G. Rickover (1957) informs the nation that it is an urgent matter of national security to raise educational standards and implement an educational tracking system to provide a *specific* kind of education for students of *specific* ability levels (e.g., college bound, general, slow), with the most talented teachers being assigned to the college-bound track. Thus, it quickly becomes natural within the American public education system to categorize students and to assign greater societal value to one group of students (gifted and/or college bound) over the others. Given this naturalized way of talking about students, it seems rather predictable that a parallel system of education—special education—will arise as an apparatus through which to disseminate statements about students who display difficulties in learning.

The socially efficient system of ranking and ordering students takes on even greater significance in light of the desegregation efforts of the mid-1960s (Tyack, 1974). As previously described, the War on Poverty, initiated during the early 1960s, targets the culturally deprived as a "special" population needing "special" programs—opening yet another space for discourse to materialize around the idea of providing a *different* education for *some* children—the basis of which eventually undergirds the rationale for expanding special education. With the desegregation of schools, Sleeter (1987) contends that "minority children were seen as behind and resegregated within the schools in special programs which helped retain white privilege" (p. 219). Given the naturalized idea that students can and should be sorted within the ranks of bright, average, and slow, it is assumed that a considerable segment of the school population (primarily children of low social class and nondominant culture) necessarily will be unable to meet "average" educational standards. To explain this "necessary" failure, educators locate deficiencies *within* these children and/or their home environments by identifying four student categories: the mentally retarded, the slow learner, the emotionally disturbed, and the culturally deprived (Sleeter, 1987). These categories serve to justify the school failure of minority children from "culturally deprived" environments, but leave unexplained why some middle-class White children are also unable to keep pace with the increased academic standards. In considering the social, political, and cultural milieu of the early 1960s, Sleeter (1987) argues that

> learning disabilities was created to explain the failure of children to meet those standards when existing explanations based on mental, emotional, or cultural deficiency did not seem to fit. Learning disabilities seemed to explain white middle class children particularly well because it did not level blame on their home or neighborhood environment, it upheld their intellectual normalcy, and it suggested hope for a cure and for their eventual ability to attain relatively higher status occupations than other low achievers. (p. 231)

Thus, we might conceptualize the field of learning disabilities as having emerged out of a larger and ongoing educational discourse about the reasons for school failure in contrast to the conventional view that the field emerged solely from the scientific discovery of an identifiable and treatable childhood pathology.

Perhaps a more accurate rendering of the birth of the learning disabilities field is one that accounts for the *interaction* between the discourse of the day and the newly available scientific information. Sleeter (1987) explains that White middle-class parents of the early 1960s, drawing upon medical research, offer educators an explanation for their failing children, namely, the existence of a biological condition recently coined a "learning disability." This explanation takes on further significance if we consider that the students served within the learning disabilities category between 1963 and 1973 were, in fact, mostly White and middle class or above (Sleeter, 1986). Thus, the demographics of students served by the earliest learning disability programs support Varenne and McDermott's (1998) contention that

> to the very extent that "merit" as measured by the evaluation of individuals is constitutive of a success *officially* recognized as success, the ideological imperative that drove Dewey and Thorndike must also lead, at the level of familial politics, to a desperate struggle to make one's own children appear legitimately better. (p. 108)

The Discourse of Normalcy and the Rise of Testing

Given that the field of learning disabilities emerged from medicine and psychology, it is no surprise that the field's methods of classification reflect a medical model approach to disability. To categorize an individual as having a learning disability is to compare that individual with others deemed as disability free and to document the specific areas of deficit inherent within that individual on the basis of that comparison. A learning disability necessarily becomes an occasion for constructing notions about normal/abnormal. As individual psycho-educational evaluations are central to the institution of special education, it is relevant to explore how "a discourse of normalcy" arose along with its subsequent contribution to the rise of testing within public education.

Davis (1997) contends that our current understanding of the word "normal," established within the English lexicon in the mid-1840s, can be traced to the nineteenth century's budding field of statistics. Initially conceived in Europe as a way to amass data regarding industrial production and public health, statistical usage is extended by Adolphe Quetelet, a French statistician, to include the notion that human characteristics (e.g., height and weight) can be tallied and averaged to construct an abstraction of a country's average or "normal" man. In constructing an ideal man, Quetelet creates the first framework within which to compare human beings as either "normal" or "not normal."

Sir Francis Galton, building upon his cousin Charles Darwin's theory of natural selection, studies individual differences in light of their significant contribution to human evolution (Hanson, 1993). Conducting his work in the second half of the nineteenth century, Galton reconceptualizes Quetelet's error curve as the normal curve of distribution, adding quartile divisions for ranking traits as average, inferior, or superior (Davis, 1997; Hanson, 1993). It is noteworthy that Galton's notion of a "normal curve of distribution" is used in current-day assessment and eligibility criteria for special education services.

Galton was also one of a number of European statisticians who drew a connection between the statistical study of human traits and the possibility of enhancing the traits of a population by disallowing its weakest members to mix with desirable members—a popular ideology of the day known as "eugenics" (Davis, 1997; Thomas & Loxley, 2001). For example, in an 1869 publication, Galton describes intelligence as an inheritable trait distributed unequally among human beings according to class and race, promoting the notion that intelligent people should marry one another to offset the increasing birthrate of the undesirable lower classes (Hanson, 1993).

Given the significance that the eugenics movements attributed to the trait of intelligence, it is not surprising that the field of psychometrics or mental measurement soon emerges. In response to a request from the French Ministry of Education in the early 1900s, Alfred Binet constructs the first intelligence test, but not without caveats about the use of mental quotients (Thomas & Loxley, 2001). Despite Binet's concerns about widespread use of his intelligence test, demand for the test immediately grows, particularly in the United States where it appeals to psychologists steeped in the positivist tradition. In 1916, Lewis Terman, a psychologist at Stanford University, modifies, expands, and renames the test (Stanford-Binet), popularizing its use in the United States (Hanson, 1993).

Not everyone embraces the arrival of the intelligence test in America. In 1922, the political scientist Walter Lippman publishes a series of articles about testers who ascribe to the notion of IQ inheritability, noting that "intelligence testing in the hands of men who hold this dogma could not but lead to an intellectual caste system" (Thomas & Loxley, 2001). Lippman's warnings prove prophetic. The intelligence scale, in the hands of American psychologists, "became not just a diagnostic device, but a powerful tool by which society could be regulated" (Kliebard, 1995, p. 93).

During the early decades of the twentieth century, proponents of the social efficiency movement believe that "scientific rationality and technology would save the modern school" (Rousmaniere, 1997, p. 3) and eliminate wastefulness by educating each class of individuals according to their *predicted* social and vocational roles. The emerging field of mental measurement provides the tools to carry out the differentiation of curriculum commensurate with measured student IQ. As Finney, a prominent sociologist of education and a proponent of social efficiency during the 1920s, contends, "IQs below 99 are not likely to secrete cogitations of any great social fruitfulness" (Finney, 1928, p. 388). In fact, Finney suggests a "leadership" curriculum for capable students and a "followership" curriculum for incapable students (Kliebard, 1995).

With scientific tools in hand, psychologists wield authority to sort students into their appropriate slots in school and ultimately within society. Given that IQ tests are developed within the scientific tradition, the legitimacy of such tools is not called into question nor is the practice of separating children on the basis of IQ scores. It is important to consider that science, in early twentieth century America, reigns sovereign as the Discourse of Truth, the source of all reliable knowledge. Therefore, what naturalizes these ideas and gives them momentum is their association with the methods of natural science (Thomas & Loxley, 2001).

The legacy of positivistic science permeates present-day special education practice. Testing sits center stage within the special education apparatus. For a child to receive any special education service, an IQ test must be administered. As such, IQ functions as the gatekeeper to special education services and ultimately to life chances. A learning disability is most often defined as a discrepancy between IQ and achievement. Thus, a learning disability, by its very definition, represents the *quantification* of ability and achievement. Although the legitimacy and reliability of IQ tests have repeatedly been called into question (Kinchloe, Steinberg, & Gresson, 1997; Kinchloe, Steinberg, & Villaverde, 1999), special education continues to revolve around Galton's nineteenth-century conception of the "curve of normal distribution" and its accompanying notions of normal/abnormal.

Special education remains mired in the methods of natural science, despite the fact that "human beings are not, like the objects of natural science, things which do not understand themselves" (Joynson, 1974, p. 2). What are the consequences of our persistent allegiance to a positivistic tradition? Thomas and Loxley (2001) suggest that

> it has involved an intentional casting off of certain kinds of knowledge—the knowledge we have of other human beings which comes by virtue of our own membership of the human species—in the assumption that these kinds of knowledge would contaminate a dispassionate, disinterested understanding of others. And in doing this, a strange kind of professional and academic language has been encouraged. Straightforward understandings have often been puffed up into something to look impressive and "scientific." (p. 29)

Special Education: A Discourse between Parents and Professionals

Special education law ensures that parents have the right to be informed, the right to challenge, and the right to appeal (IDEA, 1990, 1997, 2004). Thus, parents have the right, by law, to engage with professionals in the special education process. If, as Thomas and Loxley (2001) suggest, school personnel (e.g. school psychologists, special education administrators, special education teachers, speech language pathologists, general education teachers) adopt a dispassionate, disinterested understanding in keeping with a positivistic science tradition, how might such positioning impact the *quality* of discourse between school professionals and parents? I address this literature in two sections—the years shortly after the passage of PL 94–142 and subsequent years to the present.

Early Implementation of the Law: 1978–1985

In light of the unprecedented emphasis upon parental involvement required by PL 94–142, educational researchers in the late 1970s and early 1980s study the parent/professional collaboration as required under the guidelines of the new law. Results of these studies, however, reveal passive participation on the part of parents. For example, in a series of interviews conducted by Hoff, Fenton, Yoshida, and Kaufman (1978), parents communicate inaccurate perceptions of team decisions and express confusion and/or lack of knowledge about their rights under the law. Similarly, in observations of individual education plan (IEP) conferences, Goldstein, Strickland, Turnbull, and Curry (1980) consistently note unclear explanations of psychological testing, lack of opportunity for parents to ask questions, and the presentation of preprepared IEPs for parents to sign. Moreover, Poland, Thurlow, Ysseldyke, and Mirkin (1982) report that 79 percent of 100 directors of special education acknowledged through surveys that team meetings for the purpose of discussing eligibility and placement were held *before* meeting with the parents. These directors confirm that IEPs were mostly completed without parental input.

And what of the parents' right to challenge as guaranteed under the law? Sonnenschein (1981) and Lipsky (1985) claim that the response of professionals to parents who challenge is to label the parents in negative terms. Sonnenschein (1981) explains that

> parents who disagree with a diagnosis or seek a second opinion are "denying"; those who refuse the kind of treatment that is suggested are "resistant"; and those who are convinced that something is wrong with their child despite inconclusive tests are "anxious." (p. 64)

Thus, the implication is that parents are too emotional and subjective to meaningfully participate in educational decision making, despite their right to engage in the process.

This literature, describing parent/professional collaboration in the years shortly after the passage of PL 94–142, corresponds with my own observations and experiences in special education during those years. Certainly, a period of transition was to be expected as school districts integrated the requirements of the new and revolutionary law. Does the literature of more recent years, then, provide evidence of increased quality of collaboration between professionals and parents?

Implementation of the Law: 1986 to the Present

The literature continues to present a picture of less than optimal parental participation in the special education decision-making process. For example, in a survey of over 100 parents with children having learning disabilities, Malekoff, Johnson, and Klappersack (1991) report that "over half of the parents indicated that they were initially confused" about their child's diagnosis and described the test results and recommendations as "not helpful" (p. 420). Testimonies of more than 400 parents/family members of children with disabilities heard by the National Council on Disability (1995) confirm that parents continue to be presented with preprepared IEPs, thereby feeling "largely left out of the process" (p. 11). Engel (1993), an attorney and father of a child with a mild disability, explains that

> these professional members of the CSE [Committee of Special Education] use language to discuss disabilities that is foreign tomost parents...Although it is their child who is being discussed, it does not *seem* like their child...Their own knowledge systems seem trivial and "unscientific" in comparison to the knowledge systems of professionals. The parents sense that their comments carry less weight in the CSE meeting because they are not couched in the language of the professionals and because the parents' close relationship to their child makes them seem overly subjective. (p. 800)

Reflective of recent research trends that focus upon the impact of gender, race, and class in American schools, Harry (1992) studies the experience of low-income Puerto Rican American families within the system of special education. Results of ethnographic interviews and observations indicate minimal parental involvement in decision making and misunderstandings about educational diagnoses and placements. Parents report great difficulty understanding special education jargon and coping with the sheer volume of papers required for special education placement. Harry (1992) observes these parents deferring to authority out of cultural respect, but engaging in "a kind of passive resistance in hope of demonstrating their disapproval" (p. 486).

In a three-year longitudinal study of low- to middle-income African American parents whose children were placed in preschool special education, Harry, Allen, and McLaughlin (1995) observe IEP conferences and confirm that "parents' participation... usually consisted of listening, perhaps asking a question (usually regarding logistical issues such as transportation), and signing papers" (p. 371). Moreover, professionals state that they do not think "parents really understood much of it, but simply signed anyway" (Harry, Allen, & McLaughlin, 1995, p. 371). It is noteworthy that these professionals use educational jargon with parents, allot less than 30 minutes for conferences, and focus on completion of paperwork. Over the three years, parents become disillusioned and levels of participation decrease.

Throughout my career, I have consistently observed that more mothers than fathers or sets of parents engage in conversations with school personnel. It is not uncommon for mothers who voice instinctual concerns about their children's learning and behavior to be dismissed by professionals. For example, in their handbook for school counselors, Baumberger and Harper (1999) state that "it has been noted in the literature and in the authors' experience that mothers of children with learning disabilities tend to be overprotective, overinvolved, and even enmeshed" (p. 63). Note that it is the *mother* who is singled out as overprotective, overinvolved, and enmeshed, not fathers or parents as a unit. It appears that mothers, by virtue of their motherhood, must also contend with gender stereotypes in communicating with school personnel.

In considering the existing literature on parent/professional collaboration under IDEA, it is noteworthy that not only is there little research available, but it (excepting the work of Harry, 1992, and Harry et al., 1995) primarily reflects information gathered *about* parents from data collected through methods such as large-scale surveys (completed by special education administrators) and third-party observations of special education meetings. It is of particular interest that the researchers overwhelmingly conclude that parents are positioned outside, rather than within the discourse; yet, parents by virtue of the research designs (e.g., surveys and observations) are likewise excluded from contributing meaningfully to the conversations about them.

Discursive Practices of Special Education

How might we begin to understand the process by which professionals position themselves as authorities? How do parents become positioned as "less than" within the discursive practices of special education? If we consider the relationship of parents and American schools in general, an unmistakable pattern emerges, evident from the inception of public school through the present, that points to continual efforts on the part of schools to maintain professional dominance—primarily on the basis of "scientific" knowledge. This historical look at American home/school relationships further reveals that the degree of paternalism directed at parents is relative to their particular ethnic and cultural background, class, and/or race (Cutler, 2000; Valle & Reid, 2001).

If we locate IDEA guidelines within the historical relationship that parents have had with public schools, it is clear that the parental rights guaranteed under the law represent a significant *anomaly* in the context of public schools (Skrtic, 1995). It is of little wonder, then, that school personnel implement the law in a way that is compatible with traditional patterns of parent/professional interactions.

Power and Parent Status

To more fully locate parents within the discursive practices of special education, it is necessary to consider special education, with its roots in medicine and psychology, as a human science institutionalized within the context of public schooling. What might be the consequences of such an arrangement? In an effort to better understand the intimate intersection between public schooling and human science, I turn to the French philosopher Michel Foucault and his analyses of the human sciences as a lens through which to consider the apparatus that is special education.

Foucault (1983) describes the focus of his intellectual endeavor as creating a "history of the different modes by which, in our culture, human beings are made subjects" (p. 208). Thus, the nature of his work necessarily focuses upon analyses of the human sciences. In identifying the significance of the late eighteenth-century introduction of "the individual" (e.g., the madman, the sexual deviant, and the criminal) and the subsequent practices created and enacted around such individuals, Foucault constructs a methodology for the study of human beings that also serves to illuminate social practices of the present. His method centers on an *analysis of discourse* within the human sciences; thus, Foucault perceives the subject as a function of discourse (Dreyfus & Rabinow, 1983). Foucault explains that

> Western man [*sic*] could constitute himself in his own eyes as an object of science, he grasped himself within his language, and gave himself, in himself and by himself, a discursive existence, only in the opening created by his own elimination. (1973, p. 197)

It is this space of elimination that allowed for the objectification of human beings into subjects. A central issue, then, for Foucault (1978) is to identify how this object of discourse (e.g. madness, sex, criminality) is *made into* "an overall discursive fact" (p. 11). In an effort to unmask the production of discursive facts, Foucault (1978) poses questions such as "who does the speaking, the positions and viewpoints from which they speak, the institutions which prompt people to speak about it and which store and distribute the things that are said" (p. 11). Therefore, what interests Foucault is not the truth of what is said, but rather the apparatus itself through which what is said *can* be said in a particular society. Foucault contends that discourse can be understood as a system that "unifies the whole system of practices [within which] various social, political, economic, technological, and pedagogical factors come together and function in a coherent way" (Dreyfus & Rabinow, 1983, p. 65).

If we understand special education as a discourse through which a whole system of practices functions in a coherent way, the apparatus that produces and maintains the notion of learning disabilities becomes clearer. The parent who enters a special education committee meeting also enters a complex network in which a discourse of disability is circulated. Special education professionals, speaking from a "scientific" perspective, claim authority on the basis of the particularly Western conception of science as an objective, indisputable truth. How might a mother enter such a discourse? Foucault (1977) asks us to consider

> what types of knowledge do you want to disqualify in the very instant of your demand: "Is it a science"? Which speaking, discoursing subjects—which subjects of experience and knowledge—do you then want to "diminish" when you say: "I who conduct this discourse am conducting a scientific discourse, and I am a scientist"? (pp. 84–85)

Foucault's questions appear relevant to the discussion at hand. School professionals, speaking with the authority that science allows them, disqualify (intentionally or unintentionally) the knowledge that parents bring about their children. Is it not paradoxical that special education discourse—a system of practices in which parent/professional collaboration is legally mandated—simultaneously operates out of a scientific framework that, by its very nature, gives authority to professionals?

In his analyses of the human sciences, Foucault (1977) explores the relationship between power and science—a particular type of knowledge highly valued in our culture—as well as ways in which societies privilege one discourse over another. We might further understand Foucault's claims by considering the Russian linguist Mikhail Bakhtin's (1981) notion of addressivity within the context of special education committee meetings. Bakhtin (1986), like Foucault, recognizes the privilege given to the monologic, acontextual thinking that typifies the human sciences. Bakhtin (1981) is likewise interested in who does the speaking, the positions and viewpoints from which they speak, and the context in which the speaking occurs. If we consider the context of a special education committee meeting, it is clear who does the speaking and from what viewpoint. School personnel sit at the center of this exchange—initiating, dominating, and terminating the discourse—that revolves around scientific explanations (e.g., IQ tests, achievement test scores, and discrepancy formulas). The parent's knowledge, in contrast, appears informal (i.e., less important) in its lack of scientific verification. Bakhtin (1981) explains that such differences in language are naturally occurring stratifications of dialects and socio-ideological languages (past and present) that exist within a nation's unitary language. This differentiated speech, which Bakhtin calls *heteroglossia*, includes the everyday language of different social groups as well as the language used by various professional groups (Vice, 1997). Bakhtin (1986) refers to the spheres in which these social languages exist as *speech genres*. Speech genres, with their respective norms for language, provide the venue for a speaker's performance within a particular genre. The special education committee meeting, then, can be understood as a discourse arena in which heteroglossia is present in the differentiated speech genres of the parent and the professional. Bakhtin (1981) observes that

> many people who have an excellent command of language often feel quite helpless in certain spheres of communication because they do not have a practical command of the generic forms used in the given spheres. (p. 80)

Bakhtin refers to this insider language as the *password*—the command of which enables one to function adequately within a particular speech genre (Casey, 1993). Without access to the professional speech genre of special education, parents enter the discourse in an unequal position. Such positioning clearly intends to elicit allegiance to professional opinion rather than facilitate meaningful dialogue.

The discursive practices of special education revolve around scientific data (e.g., IQ tests, achievement tests, observations, teacher reports, speech/language evaluations, checklists) that have been gathered about the child. Such documentation and record keeping are naturalized practices of the human sciences. As Foucault (1997) contends, "The sciences of man were born at the moment when procedures of surveillance and record-taking of individuals were established" (p. 74). Of particular interest to Foucault is who is in the position of surveillance and record taking and who is the object of such surveillance and records. Foucault (1975) explains that

> this turning of real lives into writing... functions as a procedure of objectification and subjection... The examination as the fixing, at once ritual and "scientific," of individual differences, as the pinning down of each individual in his [*sic*] own particularity... clearly indicates the appearance of a new modality of power in which each individual receives as his status his own individuality, and in which he is linked by his status to the features, the measurements, the gaps, the "marks" that characterize him and make him a "case." (p. 192)

In this way, a link to the "medical bipolarity of the normal and the pathological" is unmistakable within the human sciences (Foucault, 1973, p. 35). The classification of human characteristics (and by extension human beings) as *normal* or *abnormal* takes on great significance in the human sciences. In the case of special education, the norm is the axis around which special education classification spins. To be classified as LD is to be deviant from the norm. In the tradition of human science, special education embraces a complex system of scientific, objective measures for judging what is and is not considered within the *normal* range of learning and behavior. Foucault (1975) regards such methods of normalization as a "great instrument of power" (p. 184). Within the context of special education, we must consider the "great instrument of power" that school psychologists and educational evaluators wield in their capacity to deem a child *normal* or *abnormal.* Within the structure of IDEA, the significance of such power is acknowledged through the guarantee of due process for parents. Although this guarantee exists, the onus is upon the parent to disprove professional opinion in terms of the knowledge base that is valued by school professionals, as opposed to presenting arguments rooted in other "ways of knowing."

Foucault (1977) suggests that "it is not possible for power to be exercised without knowledge" and that "the exercise of power perpetually creates knowledge and, conversely, knowledge constantly induces effects of power" (p. 52). In other words, as the concept of learning disabilities entered our language, objects of knowledge emerged along with an ever increasing body of knowledge about learning disabilities. Simultaneously, *power* surfaced alongside *knowledge*, creating a continual interplay that Foucault (1977) describes as "the articulation of each on the other" (p. 39). Foucault (1978) is careful, however, to differentiate his notion of power from the conventional idea of power as superimposed; instead, power is conceived as "produced from one moment to the next, at every point, or rather in every relation from one point to another" (p. 93). In considering

Foucault's ideas about the relationship between power and knowledge, might it be that mothers both participate in and resist disqualification of their own knowledge?

Bakhtin describes such dialogic tension as a "constant struggle between the centripetal forces that seek to close the world in system and the centrifugal forces that battle completedness to keep the world open to becoming" (Clark & Holquist, 1984, pp. 79–80). It is the intersection and interaction of speech genres (e.g., authoritative discourse and internally persuasive discourse—discourse that is "denied all privilege, backed up by no authority at all, and frequently not even acknowledged in society") that interests Bakhtin (1981, p. 342). Certainly, the tension between authoritative discourse and internally persuasive discourse influences the "degree to which one voice has the authority to come into contact with and interanimate the other" (Wertsch, 1991, p. 78). Nevertheless, it is internally persuasive discourse that is dialogic, that enables people to go beyond internalizing dogma to infuse the message into their own understanding and collaborate in the construction of true communication—the type of parent/professional collaboration envisioned within the spirit of the law.

Power and Parent Gender

Given that the overwhelming majority of parents who participate in the special education decision-making process are mothers, it appears relevant to consider the role of gendered knowledge and embodiment in the apparent power disparities between parents and special education professionals. Over the last few decades, human science methodology and assessment practices have been increasingly critiqued for privileging monologic worldviews within scientifically produced knowledge. Feminist scholars contribute to this critique by exposing masculine biases inherent in producing scientific knowledge (Olesen, in Denzin & Lincoln, 1998). Thus, feminist assertions about gendered knowledge and the effects of embodiment upon knowledge production (or suppression) seem relevant to understanding the power disparity between mothers and special education professionals.

In her critique of the human sciences, Grosz (1993) asserts that objectivist inquiry, with its emphasis upon detached, neutral methodologies grounded in reason, produces masculine knowledge believed to represent universal Truth. Under the guise of neutrality and reason, Grosz asserts that men "evacuated their own specific forms of corporeality and repressed all traces of their sexual specificity from the knowledges they produce" (p. 204), thereby effectively absolving the subject of knowledge from accountability in the consequences of knowledge production. Grosz's critique of objectivist inquiry is useful in understanding the inherent power accorded to psycho-educational reports. Test scores, generated through supposedly neutral, unbiased methodology, represent Truth, thereby releasing the school psychologist from responsibility for knowledge production and its consequences. Test scores, commanding center stage, stand alone in the representation of Truth, resisting and suppressing argument about their validity as Truth.

Grosz (1993) further contends that masculine knowledge "remains unrecognized as such because there is no other knowledge with which it can be contrasted" (p. 204). If we understand psycho-educational evaluations to represent masculine knowledge, it is not surprising that school personnel routinely dismiss mothers and their knowledge assertions. In contrast to the privileged, uncontested ("masculine") knowledge of test scores, mothers' ("feminine") knowledge appears subjective, emotional, unworthy, and irrelevant to valid knowledge production about their children. Thus, school personnel perceive mothers as eager recipients of privileged knowledge, rather than active producers and contributors of knowledge. Should a mother challenge masculine knowledge claims, school personnel typically dismiss the mother as "being in denial"—given her irrational, emotional, and overly subjective, decidedly female perspective, the antithesis of all that masculine knowledge purports to be.

Echoing Grosz's concerns about *whose* subjectivity is valued in knowledge production and the power embedded within knowledge production, Shildrick (1997) examines additional questions about subjectivity. What is the positionality of the knowledge agent? Who is allowed to speak? How are debates resolved about universality versus relativity? Shildrick challenges the notion that the answer to such questions lies in either one of two extremes—the claim of a feminine subjectivity in *sameness* to the masculine subjectivity or the claim of a feminine subjectivity in *opposition* to masculine subjectivity. The claim of sameness reinforces masculine subjectivity as ideal and normative, leaving intact one mode of thought to be characterized as rational; such a stance does not acknowledge that knowledge is partial and located. Claims of opposition lead to the same result because, through contrast, they reiterate "the very qualities which have disqualified not just women, but all the others," thereby perpetuating the subjugation of feminine voice and the possibility for feminine subjectivity (Shildrick, 1997, p. 156). She further argues for "acknowledgement of differences as expressed in the specificity and multiplicity of both sexes . . . the task [being] to recognize all the others and acknowledge their inclusiveness in the construction of the gendered self" (Shildrick, 1997, p. 157). What Shildrick's ideas might offer to special education is a conceptualization of knowledge as partial and located. Grosz (1993) reaches a similar conclusion in her call for acknowledgment of the perspectival nature of knowledge, arguing that the current critique of human sciences is perhaps better understood as a "crisis of specificity"—a so-called crisis that acknowledges the particularities of knowledge and exposes knowledge production as an interested, located activity. We might understand special education practice as *needing* a "crisis of specificity" because of its present refusal to embrace knowledge beyond that which is gleaned from objective inquiry.

In her view of knowledge production as an act of power, Grosz (1993) points out that knowledge requires the interaction of power and bodies to ensure allegiance to the body of knowledge created. If we consider bodies in connection with knowledge production, it is necessary to acknowledge the implication of bodies that are "sexually coded" by society. Grosz (1993) explains that "sexual differences demand social representation insofar as social roles and procreative functions are not governed by instincts or 'nature' but are socially required, produced, and regulated" (p. 202). In exposing the universal as masculine knowledge, questions inevitably emerge about the consequences of holding women—who by virtue of their sexually coded bodies not only experience the world differently than men but also with far less privilege than afforded to men—to an "ideal" male norm. What questions might we pose about the consequences of interpreting a child's educational needs solely through the perspective of *masculine* knowledge? What are the educational consequences of dismissing a mother's particular way of knowing her child?

Young (1990) shares Grosz's interest in the relationship between embodiment and knowledge production, arguing that men and women experience the world differently by virtue of their embodiment. For example, Young argues that patriarchal logic is implied in the binary of asexual mother/sexualized beauty, asserting that "patriarchy depends on this border between motherhood and sexuality" (p. 197). Thus, mothers are viewed as self-sacrificing, pure, domesticated, and committed to maternal duties. If we return to the scenario of the special education committee meeting, we see how mothers become defined within the confines of a patriarchal definition of motherhood. That is, mothers, by virtue of motherhood, are viewed as self-sacrificing, dedicated, overprotective, and enmeshed with their children. Therefore, mothers can be disqualified, on the basis of their motherhood, from the rational and objective knowledge production privileged by special education professionals. The influence of gender stereotyping upon educational decision making becomes clearer when a father attends an eligibility and placement committee meeting. Fathers typically are afforded a greater role in discourse and greater validation of their knowledge contributions. Moreover, fathers appear generally exempt from the assumptions about subjectivity that plague mothers.

Might this not be another example of Young's (1990) assertion that men and women experience different ways of being in the world by virtue of their embodiment?

In proposing a postmodern feminist ethic, Shildrick (1997) dismantles such Western dichotomies of mind and body with respective association with the masculine and the feminine. However, she insists that a postmodern feminist ethic does not suggest a disembodied subject as reflected in masculine subjectivity. On the contrary, the "feminist rewriting of the subject demands an attention to the corporeal body" (Shildrick, 1997, p. 167). Therefore, unity of mind and body is accomplished through the affirming embodiment; yet, this affirmation must neither essentialize gender differences nor deny differences resulting from the experience of living in a sexed body. Shildrick (1997) strives to "integrate the excluded without losing touch with the specificity" (p. 172), thereby promoting openness toward differences without privileging one subjectivity over another.

Having attended innumerable special education committee meetings, I am struck by the overwhelming presence of women in this decision-making process. In addition to the preponderance of mothers, the professionals (e.g., teachers, principals, school psychologists, guidance counselors, social workers) are also predominantly female, and most are also mothers. It is interesting to me that professional women/mothers present Truth about a child in the spirit of paternalistic benevolence. For these professionals to believe that what they are doing is in the best interest of others and to justify the dismissal of the mother's perspective, they must choose to recognize themselves only as professionals and not *also* as women/mothers. What is notable is the seeming lack of awareness on the part of professionals both about the pain they may inflict as well as the sense of betrayal that mothers experience at the hands of women/mother professionals. Yet, such puzzling participation of women in the perpetuation of patriarchy is well documented. Feminist historian Gerda Lerner (1986), for example, explores this phenomenon within her study of the creation of patriarchy.

> What could explain women's historical "complicity" in upholding the patriarchal system that subordinated them and in transmitting that system generation after generation, to their children of both sexes? (p. 6)

If we return to the scenario of the special education committee meeting, we can see that women/mother professionals, operating out of a patriarchal stance, claim *sameness* to the masculine subjectivity inherent in scientific inquiry, while mothers claim *difference* to the masculine subjectivity by challenging scientific inquiry with the particularities of ways in which mothers know their children. Women/mother professionals, performing the masculine role, allow mothers access into the decision-making arena (in this case, required by law), yet require them to confirm or disprove Truth that *already exists* within the scientific data presented—effectively replicating the routine positioning of women within our culture. Lerner (1986) explains how such positioning operates.

> The women finally, after considerable struggle, win the right of access to equal role assignment, but first they must "qualify." The terms of their "qualifications" are again set by the men; men are the judges of how women measure up; men grant or deny admission. They give preference to docile women and to those who fit their job-description accurately. Men punish, by ridicule, exclusion, or ostracism, any woman who assumes the right to interpret her own role or—worst of all—the right to rewrite the script. (pp. 12–13)

Lerner's explanation neatly parallels the dynamics that can occur within special education committee meetings when women/mother professionals perform a masculine role. For example, IDEA guarantees mothers the right to participate in educational decision making, yet "men" (i.e., women/mother professionals who perform a masculine role) set the terms of participation by privileging scientifically produced knowledge over other ways of knowing. On the basis of my lived experiences within the institution of special education and the literature review regarding parent/professional relationships under IDEA, it appears that these "men" indeed prefer "docile women" who fit their "job-description" for mothers (e.g., passive and accepting). On the other hand, mothers who insist upon equality and assert their opinions (i.e., "assume the right to interpret her own role and the right to rewrite the script") disrupt convention, prompting these "men" to label them as irrational, overly subjective, and incapable of participating meaningfully.

Given this either/or subjective stance of professionals and mothers, it is unsurprising that the special education committee meeting can become a battleground—with either side refusing to acknowledge the knowledge position of the other. Moreover, it is relevant to consider Shildrick's (1997) notion that a claim of sameness or a claim of difference to a masculine subjectivity perpetuates the subjugation of feminine voice. Within this kind of discourse structure, there is little space for the kind of collaboration envisioned in IDEA.

Power and Parent Race/Class

It is impossible, of course, to consider the institution of special education apart from the system in which it is embedded. Its history is inextricably intertwined with the history of public schooling. As such, the institution of special education operates under similar assumptions to those that undergird the larger system of American public schools.

As previously established, public schools have a long held a paternalistic attitude toward parents. From the early part of the twentieth century to the present, leaders of the dominant culture have recognized public schools as a viable and effective avenue for disseminating particular discursive statements, specifically those statements that support White middle- and upper-class interests and values (Kliebard, 1995; Valle & Reid, 2001). The degree of paternalism that schools have directed and continue to direct toward parents is directly related to the degree to which the parents differ from the dominant culture. Public schools have long regarded parents (especially mothers) who belong to low-income, nondominant cultures as incapable of raising their children properly (meaning like White middle-class mothers) and as legitimate sources of blame for school failure (Greene, 1978; Valle & Reid, 2001). Given this history of paternalistic treatment toward parents who fall outside White middle-class culture, how might issues of culture, class, and race contribute to the power disparities between parents and special education professionals?

I have established numerous ways in which parent/professional collaboration falls short of that envisioned by IDEA. However, IDEA's vision of collaboration appears even farther removed for low-income and culturally diverse parents (Harry, 1992; Harry, Allen, & McLaughlin, 1995; Kalyanpur & Harry, 1999; Kalyanpur, Harry, & Skrtic, 2000). In considering why low-income and culturally diverse parents may choose not to participate or to participate passively in the special education decision-making process, Kalyanpur, Harry, and Skrtic (2000) suggest that "the principle of parent participation is based on ideals that are highly valued in the dominant culture" (p. 122). In other words, the concept of parental rights embedded within special education policy may seem alien to some families, as might the particularly American values of equity and civil liberties (Bean & Thorburn, 1995; Holdsworth, 1995). To cultures that place higher value upon the needs of collective society than the individual, special education's emphasis upon the individual may seem foreign indeed (Kalyanpur & Harry, 1999).

The discursive practices of special education operate within a scientific, objective framework. Hall (1981) contends that cultures for which objectivity holds great value are "low context"—that is, cultures that rely upon decontextualization as a means to generalization. In contrast, cultures that are "high context" "accept, even encourage, conclusions that tolerate greater ambiguity" (Kalyanpur & Harry, 1999, p. 7). Thus, parents from high-context cultures who challenge the methods and interpretations of a low-context culture may be perceived as deficient and "non-compliant"—a label that, in and of itself, implies moral hegemony (Fadiman, 1997).

In light of the medical model in which learning disabilities are explained as deficits within the individual, parents who are both culturally diverse and low income may perceive the implication of a "double deficit," in the sense that not only is their *child* considered deficient so is the *culture* in which they are raising their child (Kalyanpur, Harry, & Skrtic, 2000). Moreover, the conception of disability as intrinsic and treatable may be incomprehensible to parents from cultures in which disability "has spiritual causes, is temporary, is group owned, and must be accepted" (Kalyanpur & Harry, 1999, p. 45).

Much like the assertion that men and women experience different ways of being in the world by virtue of their embodiment, Delgado (1990) argues that people experience different ways of being in the world by virtue of race and racism. In considering ways in which race might contribute to the power differential within special education committee meetings, it is relevant to acknowledge the historical Western ideology of the White and Black binary (Anderson, 1998; Gould, 1996); in other words, "Whites are an intelligent, diligent, and deserving people; Blacks are simple, lazy, and undeserving people" (Tate, 1996, p. 200). Sleeter (1987) draws upon this ideology to support her argument that the category of learning disabilities was, in actuality, conceived as a way to explain the failure of White, middle-class children (expected to achieve by virtue of being White) in contrast to the *expectation of failure* in minority children whose cultures were considered deprived in relationship to White, middle-class culture.

Let us consider the relationship between race and the assessment practices that drive special education. Padilla and Lindholm (1996) claim that IQ tests support a racial inferiority paradigm in ways that seem natural. According to them

> (a) the White middle-class American (often male) serves as the standard against which other groups are compared;
> (b) the instruments used to measure differences are universally applied across all groups, with perhaps slight adjustments for culturally diverse populations; and
> (c) although we need to recognize sources of potential variance such as social class, gender, cultural orientation, and proficiency in English, these factors are viewed as extraneous and can later be ignored. (p. 199)

Is it any wonder that a non-White mother might not recognize her child in test results based upon White middle-class experience?

An Intersectional Framework

In referring to doctrinal and political discourses, Crenshaw (1993) argues that it is counterproductive to conceive of multiple systems of subordination (e.g., gender, race, class) as separate entities. For example, Crenshaw (1993) asks, "How does the fact that women of color are simultaneously situated within at least two groups that are subjected to broad societal subordination bear upon problems traditionally viewed as monocausal—that is, gender discrimination or race discrimination" (p. 114). I offer a similar line of thinking toward a more comprehensive and textured understanding of parent/professional collaboration under IDEA. For the purpose of clarity, I chose to present separate arguments around the impact of race, class, and gender upon parent/professional collaboration; however, I believe that this phenomenon can best be understood within an intersectional framework that simultaneously takes into account the multiple systems of subordination that mothers must negotiate.

Conclusion

A complex historical relationship exists between parents and public schools. In that public schools embody the larger cultural and political discourses of society, it is relevant to consider the role of such discourses in negotiations between parents and schools in general—and negotiations between parents and special education personnel in particular. Given that historical evidence suggests a consistent pattern on the part of public school professionals in exerting dominance over parents (especially parents who differ from the dominant culture), the collaboration envisioned between parents and professionals under IDEA can be considered to be an anomalous institution within the culture of public schools. In light of this lack of precedence in regard to parental rights, it is little wonder that the special education literature confirms that parents have yet to be accepted as authentic collaborators under IDEA guidelines. When parents step into the arena of special education discourse, they enter an already ongoing drama in which the principal players speak the elaborate language of science and law and offer walk-on roles to parents. Additionally, the literature suggests that the positioning of parents in this ongoing drama is conflated by their race, class, and gender.

Chapter Three
The Early Years: First Generation Mothers

I am defining this first historical time frame as the era before and immediately following the 1975 passage of the Education for All Handicapped Children Act (PL 94–142). In this chapter, I document the experiences of five mothers whose children, born in the 1960s or 1970s, were of school age during this period and labeled learning disabled (LD): Della whose daughter Jennifer was born in 1960; Cam whose daughter, Pat, was born in 1967; Mimi whose son Pete was born in 1968; Cosette who has two daughters, Katie born in 1974 and Lily born in 1977; and Lindsey whose son Charlie was also born in 1977 (see appendix B). All of the mothers are White and middle to upper class (see appendix C), reflecting the typical demographic for students labeled LD during this period (Sleeter, 1987). The children of Della, Cam, and Mimi, born in the 1960s, attended public school before, during, and after the implementation of PL 94–142, while the children of Cosette and Lindsey, born in the 1970s, entered school with PL 94–142 newly in place.

Participant Snapshots

I begin this chapter with participant snapshots, or brief biographical sketches, to acknowledge each mother as an individual with a particular story to tell as well as to orient the reader to the general positionality of each participant. I also describe the nature of my relationship with each mother. Reflecting upon participant representation in text, Lincoln (in Tierney & Lincoln, 1997) states, "Multiple stories feed into any text; but, equally important, multiple selves feed into the writing or performance of a text, and multiple audiences find themselves connecting with the stories which are told" (p. 38). Thus, the snapshots that follow represent my efforts to convey to the reader a more intimate sense of each mother in her own right as well as to reveal the multiple selves that I bring to bear upon this work (e.g., teacher, researcher, relative, consultant, friend). It is not without trepidation that I present these snapshots. I heed Lincoln's caution to researchers who collect and publicly present the stories of "silenced voices" to consider how such representations might be yet another "form of dominion" (p. 44). I acknowledge, then, that I write these snapshots from within the multiple perspectives that I hold. The snapshots do not represent "the truth" about any participant. Rather, I present my own partial and situated understanding of the mothers and the stories they chose to tell.

Della

I cannot recall my childhood without thoughts of my aunt Della springing to mind. Della is and has always been a "second mother" to me. Given that her eldest daughter, Katherine, and I were inseparable during childhood, Della became my second mother by virtue of the amount of time I spent in her home, particularly during blissful summer months.

I do not remember how I knew that Katherine's younger sister, Jennifer, had problems learning in school nor can I remember a time when I was not aware of this. I do not recall anyone in my large, extended Southern family speaking about Jennifer's problems. Although the nature of her difficulties was never explained to me, I somehow knew that I was expected to make concessions to Jennifer because of her unnamed differences. I accepted this expectation without question.

Given my intimate relationship with Della's family, I witnessed firsthand my aunt's ongoing anguish regarding Jennifer's learning, social, and behavioral challenges. Over the years, I watched as she pursued every possible avenue in hopes of finding "the" answer for her child. I understood this to be Della's job as The Mother. Not unlike most fathers of the 1960s and 1970s, my uncle's role was that of The Provider. I have little doubt that my interest in mothers, in part, originates within these childhood memories.

Della's narrative traces the lonely course of a White middle-class mother in a small Southern town who struggles to find ways to educate a child who is not learning in school. Given that most of Jennifer's education takes place before the passage of PL 94–142, Della describes the challenges of mothering during a time in which the conceptualization of a learning disability does not exist. In her dogged pursuit of resources outside of the school context (e.g., tutors, psychologists, psychiatrists, and dance, piano, art, and swim teachers), Della recalls that each of these experts conveys the same message—that she, as the mother, somehow must be responsible for Jennifer's learning, social, and behavioral challenges.

When I approached my aunt about participating in the study, she readily agreed, although, I believe, not entirely without reservation. By engaging in this narrative study together, Della and I transgress the unwritten Code of White Southern Families, which requires that family members not speak aloud, sometimes even among themselves, about the troubles that exist within family life. Thus, I acknowledge Della's courage in telling her story and cherish the experience this narrative study afforded us.

Cam

In the mid-1980s, I transitioned from my position as a middle school learning disability resource teacher to a high school position as the teacher of students dually labeled as language impaired and LD. Pat, Cam's daughter, transferred into my class from another high school at the suggestion of a school psychologist who knew both the family and me.

I recall Pat, a petite teenage girl with striking blue eyes and porcelain skin, as she entered my classroom for the first time. Cam followed close behind, and I saw immediately from whom Pat had inherited her flawless complexion. I learned shortly that Cam was a top executive for an international cosmetics company. In her many years as an executive, Cam trained thousands of women in the business of sales. The heart of Cam's work, however, centered around her devotion to empowering women to believe in themselves. With the same charisma and positive energy that she brought to her work, Cam devoted herself to mothering Pat—whom she credits with having taught *her* about life—in ways that would open the doors of life to her child. In the telling of her narrative, Cam takes us into the inner world of a mother who resists professional assumptions about her child, tirelessly fighting for her daughter's right to find her own place in the world.

As Pat's former teacher, I am but one of many people who participated in the life of this now married, working woman. Cam's approach to motherhood pays true homage to the adage, "It takes a village to raise a child." In reflecting back upon Pat's life, Cam muses, "Pat has *become* to us the greatest blessing of our lives. You know, I look at her today and think what I would have missed if she had been any different from the way she is."

Mimi

As a public school special education teacher, I knew of Mimi Bing before we ever met. With her election to the local school board in the early 1980s, Mimi established a reputation for being an outspoken advocate for children with disabilities. Around this time, I was contacted by the area superintendent and informed that I would be the teacher of a newly created class for high school students dually labeled as language impaired and LD. He explained that the class had been created for Mimi's son, leaving unsaid the implications this held for me as his teacher.

Mimi and I quickly developed a strong and close relationship. As I had suspected, Mimi's reputation as demanding came out of her rightful insistence that the district follow through on the requirements of the newly implemented special education law. Once assured that I could and would meet Pete's educational needs, Mimi believed that the battle for her child's education was finally over. Reflecting upon that moment of trust between us, Mimi recalls, "I gave him to you."

Mimi's narrative might be understood as the archetypal story of a reluctant heroine who confronts life's obstacles on the way to transformation. A self-described introvert, Mimi traces the challenge of confronting experts about her child from his troubled infancy through his years of public education. While struggling to find her way, PL 94–142 emerges to provide Mimi the legal grounding for her journey. As her engagement with the new law deepens, Mimi decides to run for the local school board. She explains, "It had never, ever occurred to me to run for school board. I am a *mom* who has been able to do what I've done for my child in spite of the fact that I stutter. How can I be a public official and *stutter*? And be as *shy* as I am?" Not only is she elected to the school board, Mimi later becomes the first woman to serve as chairperson. She spends 20 years on the board, long after her son graduates from high school, working indefatigably for parents of children with special needs.

Cosette

I first met Cosette, mother of three and a registered nurse, when she enrolled her daughter Lily at the private school for children with learning disabilities where I served as director of education. A few years later, I evaluated Cosette's oldest daughter, Katie, in my position as an educational diagnostician at a developmental pediatrics clinic. Through the years, Cosette and I have maintained both a friendship and a professional relationship.

Cosette, organized, competent, and attractive, is the picture of White Southern womanhood. She married well, raised three beautiful and intelligent children, and maintains a home that rivals those featured in the pages of *Southern Living Magazine.* Yet beneath this idyll of domestic perfection, Cosette's narrative belies the intense and ongoing grief of a mother who feels emotionally abandoned by her husband in the raising of three children labeled as having learning disabilities and attention deficit hyperactivity disorder (ADHD). Parenting during a time in which learning disabilities and ADHD were just beginning to be understood, Cosette describes her consultations with specialists in vivid detail, recalling, in particular, her profound response to what feels like attacks upon her motherhood. In relating her life as a mother over two decades, Cosette returns again and again to the pervasive and overwhelming sense of isolation in every area of her life—within her marriage, her relationships with other mothers, her relationship with extended family, and her interactions with professionals.

Lindsey

During the mid-1980s, a group of parents approached me with their vision for a private school for children with learning disabilities. At that time, there were only two private schools for children labeled LD in our state, both located at a distance too far to make a daily commute feasible. Although PL 94–142 had been in place for seven years, these White middle- to upper-class parents, who described themselves as "movers and shakers" of the community, were either dissatisfied with the public school services their children received or had children who were unable to cope with the curriculum demands of the city's most prestigious private school. I accepted their offer to cofound the school and serve as its first director of education.

Lindsey was among this group of founding parents. Unhappy with the quality of learning disability services offered by the public schools, Lindsey, a certified teacher, collaborated with other parents to create an educational alternative for children labeled LD. Facing financial consequences as the result of an impending divorce, Lindsey ironically was unable to afford the tuition for her own son, but joined the new faculty as a social studies teacher. (As a faculty member, she later received a reduced tuition rate that enabled her son to attend.) Although we both eventually moved on to other professional pursuits, Lindsey and I have maintained a deep and lasting friendship and a mutual respect for one another's work in the field of learning disabilities.

Lindsey situates her narrative within the context of her upbringing in a White, Southern, upper-middle class, and Protestant "good family." She describes how she diligently followed The Rules for Good White Southern Daughters of the time by marrying in her early 20s and bearing three sons before age 30, fully expecting to reap the inevitable life rewards that would follow from living the "right life." Her impassioned narrative chronicles a journey that takes her far from the life her family expected and the life she envisioned for herself. Faced with neither a perfect marriage nor perfect children, Lindsey struggles to make sense of both the failure of her marriage and the school failure of her youngest son, Charlie. From these experiences, Lindsey learns to resist and disrupt both family and societal expectations (including those held by school professionals) for the role of The Good Mother. In doing so, she redefines the meaning of motherhood as well as the meaning of success, not only for her son, but also for herself.

Lindsey explains the reciprocal nature of her relationship with Charlie in the following excerpt from her narrative: "I used to think that he was on a path and that I was guiding him. And now I just feel like I am so fortunate to be on that journey *with* him."

First Generation Mothers: A Collective Narrative

In the following section, I present a collective narrative crafted from the mothers' individual stories that take place between the 1960s and mid-1980s. The narrative reveals their perceptions of the power distribution within special education discourse, how they position themselves within and against special education discourse, and the consequences of special education discourse in their lives. The stories are grouped by four broad themes: (1) the language of experts; (2) conflicts in shared decision making and individual education plan (IEP) implementation; (3) devalued knowledge of mothers; and (4) the influence of race/culture, social class, and gender.

The Language of Experts

In the years prior to PL 94–142, parents with financial means turned to professionals within the fields of psychology and medicine (e.g., pediatricians, family therapists, psychiatrists, speech/language pathologists, neurologists) for answers about their children's learning problems. Operating within a therapeutic culture that routinely medicalized disability, professionals relied upon the language of their respective fields, thereby framing learning difficulties in terms of "deviance from the norm, as pathological condition, as deficit" (Linton, 1998, p. 11). The field of special education, emerging from the disciplines of medicine and psychology, would adopt, rather predictably, this medicalized discourse.

Della, the eldest participant, recalls that her daughter Jennifer demonstrated difficulty learning as early as first grade. In that public schools in the 1960s did not offer, as a rule, special services for nonimpoverished and presumably "normal" children who struggle academically, Della relies upon her contacts as a public school teacher to locate an experienced reading tutor whom she could pay to teach her child. Although her reading achievement improves as a result of private tutoring, Jennifer's overall academic performance remains problematic.

> D: Jennifer learned to read, but she was still slow in most everything else. When she got to middle school, it was *really* hard. I did not find any personnel at that school that showed *any* interest in her *whats*oever. But, I still had tutors. She even went to a psychiatrist and a therapist.

40 I remember two different versions of what they felt about Jennifer. One—after she had been to the mental health place in Greenburg—said he thought she was a very *disturbed* child. So, her pediatrician recommended a psychiatrist who dealt with children . . . Then, the psychiatrist, after she saw her for—hmmm, a year or so—she said, "I can find nothing *wrong* with this child. I think she's a *happy* child. Just treat her *normally*." But, she *still* did not learn **[slight pause]**
55 normally . . . There was no allowance for individual learning or instruction **[pause]** that I saw.

Like many White middle-class mothers faced with a child "not learning" during the 1960s, Della looks to the medical community for answers about her daughter's persisting learning, social, and behavioral challenges. In stark contrast to being positioned by school personnel as invisible ("I did not find any personnel at that school that showed *any* interest in her *what*soever" [lines 37–38]), Jennifer becomes *highly* visible within a medical context as the object of what Michel Foucault (1977) refers to as "the normalizing gaze, a surveillance that makes it possible to qualify, to classify" (p. 184). Medical experts, operating from within a discourse of science, evaluate whether or not Jennifer's observable behaviors meet the criteria for "pathology" or "not pathology." Ironically, she is pronounced simultaneously by different medical experts as abnormal ("very *disturbed*" [line 42]) and normal ("nothing *wrong*" [line 53]). And, as Della points out, Jennifer "*still* did not learn" (line 54), regardless of how the experts chose to label her—normal *or* abnormal. Della pauses after making this statement, seemingly reflecting upon her evaluation of those events from the vantage point of the present. Despite her faithful engagement with the medical community, Della recalls that nothing changes in regard to Jennifer's educational progress. In fact, she describes public education before the passage of PL 94–142 as making "no allowance for individual learning or instruction" (lines 55–56). Della's narrative illustrates the lack of intersection, before the passage of PL 94–142, between discursive frameworks used by the medical community and those used by public schools for conceptualizing (or not) the learning and behavioral problems of seemingly "normal" children.

In the following account, Della describes her feelings about a medical procedure suggested by her pediatrician

D: The many medical places that we took her to, he [her husband, Dean] would take her. In fact, when she was—I guess middle school age—she had what was called then a brain wave test. I don't know what all
235 that included. And I just could not make myself go with her **[long pause]**. And Dean carried her. I guess I look back at it now and think

> "Well, I should have gone." It probably would have made her feel more comfortable [**pause**]. But, I thought, "I just can't handle that" [**voice lowers**].

Della's account reflects the medical conceptualization of learning difficulty as a pathological condition with presumed etiology in the brain, commonly diagnosed as minimal brain dysfunction (MBD) in the 1960s and early 1970s. Della recalls the tension she experienced between complying with "expert" medical advice in an effort to help her daughter (which required that she believe that a brain wave test would yield valuable information about her child) and her instinctual uneasiness about the procedure. On the day of the appointment, Della decides that she cannot bear witness to it, revealing the degree of discomfort she feels about the way medical experts act upon and talk about her child. And in the end, of course, the brain wave test yields no definitive information about Jennifer's "condition" nor any information about how she learns or how to best provide instruction. On the one hand, Della accepts being positioned by the medical discourse as The Good Mother—compliant and passive—and on the other hand, she asserts a degree of agency in her refusal to attend the medical procedure as a Good Mother would be expected to do.

Given that psycho-educational testing was not routinely offered within public schools during the 1960s, Della seeks a private evaluation for Jennifer.

> D: The only thing I remember about testing—as testing goes today—was that I had her tested at a private place in Greenburg. I don't
> 255 remember the name of the place. And I went back for a conference with him after he got the tests results. And he said that [**slight pause**] he found her to be low average, that she would make mostly Cs and Ds, Ds and Cs, maybe a B and that turned out to be true [**slight pause**]. I mean she made Bs in um . . . now she took some kind of
> 260 Home Ec course and she made As. And in Art, she made good. But all the academics [**pause**], if she made a D, she was doing well.

In Della's recollection of this feedback conference, the psychologist pronounces Jennifer to be "low average" (line 267) in cognition, which, in turn, constructs her as doomed to being, at most, a student who "would make mostly Cs and Ds, Ds and Cs, maybe a B" (lines 257–258) and, as Della concedes, this rather bleak and presumably impervious prediction "turned out to be *true*" (line 258). Not unlike the purpose of the brain wave test to detect neurological pathology, the IQ test, in this case, is used in much the same way—as a presumably reliable and valid scientific tool that can identify (or rule out) the presence of cognitive abnormality. As

Della reports, so much faith, in fact, is placed in the Truth of the IQ score that the psychologist sees no reason to gather other sources of information about Jennifer beyond the testing context and, with confidence, predicts Jennifer's academic future as a sealed fate.

If learning problems result from presumed neurological dysfunction, thereby existing solely in the heads of children as the psychologist in Della's narrative suggests, it is of interest, then, that Della repeatedly receives another message from professionals: "I always felt like I was being the one blamed." Echoing Della's perception of being blamed by professionals, Cam vividly recalls the "language of blame" within her interactions with professionals during the 1970s.

> C: And from the *very* start, from the *very beginning* of testing, the *assumption* that I dealt with first of all was that the parents were doing
> 90 something wrong to have caused these problems. In *every*, every situation I went to, I can *never* remember *one* that I didn't have to *first* emotionally and verbally battle—"What's going on in your home? Is there dysfunction in your home? Is there violence in your home that would be causing this child to have these problems?" And as a *mother* who is already hurting
> 95 because your child is not right and who already *feels* that you've done something wrong or this child would be *fine*. You *already* feel like, you know, she was in my womb. What did I *do* that would have caused this child not to be right. Was there something *I* did during pregnancy or during delivery? So, you're already just really battling within yourself
> 100 about why has this happened and what do I need to do and trying to make sure that you deal with the guilt feelings that you feel that aren't *even* right. It's not that there is *validity* for them, it's just that they're natural because the thing you want *most* of all is for your child to be healthy and normal and have a wonderful life. So when you see this child that has
> 105 some sort of disability of some kind, you just start *digging* within yourself. *Then* you go to the professionals and the first thing they say to you is—"Well, now what are *you* doing to have caused this?" Now you've got this *huge* emotional hurdle to cope with as you sit there and try to explain that you're doing nothing. That you've given this child everything.
> 110 That you've fed this child, you've bathed this child, you've not abused this child. And that's a very difficult place. I think the mother is very *fragile* and very *vulnerable* and almost needs an advocate *herself* as to how can I *deal* with this child that I love so much and not knowing *what* to do. And then when people start bombarding you with suggestions that
> 115 it may be *your* fault, it is *really* very emotionally difficult to cope with.

Cam's narrative provides us not only with rich insight into *how* mothers become positioned by professional discourse, but also how they may *experience* such positioning. Cam recalls meeting with professionals in the context of an evaluation process for her daughter in which a power differential exists between the experts, positioned as possessing knowledge, and the mother, seeking access to their expert knowledge. In her recollection, professionals control the language of the meeting. At the word level of text, Cam frames her experience as an ongoing site of contestation, underscored by her repetition of and emphasis upon the word "every" (line 90) and her choice of words such as "battle," (line 92) "hurdle," (line 108), and "bombarding" (line 114) to describe the nature of these interactions. Moreover, she refers to meeting with professionals as something to be "dealt with" (line 89) or that she had to "cope with" (line 115). It is of interest that while Cam resists aspects of professional discourse, she engages in the binaric language available to her within a medical model of disability, describing her daughter as "not right" (line 95) in contrast to being "fine" or "normal" (line 104).

Cam's narrative suggests imagery of a tribunal in which a mother must defend herself against accusations of guilt until she can prove her innocence. She recalls being placed in a defensive position by a powerful professional discourse that assumes authority to publicly demand information about what, in most other social circumstances, would be considered personal and private. At the heart of the narrative lies Cam's intense discomfort in becoming, herself, the object of the professional community's "normalizing gaze" (Foucault, 1977). Cam's recollection of how she had to "sit there and try to explain to them that you're doing nothing" (lines 108–109) reflects positioning that feels both demeaning and powerless. Not only is her child subjected to evaluation as to whether or not she is "normal," but also her motherhood is subjected, by extension, to professional evaluation as well.

Mimi, recounting an experience remarkably similar to the one Cam describes, relates her emotional experience of being subjected to authoritative probing for personal information.

> M: Then, they sent us to a psychologist who
> did a social study on us—on Pete and me—and asked the most incredible
> *penetrating* questions you have *ever* heard in your life [**voice lowers**].
> I mean about our family, about our family relationship, about Harry and
> 1015 me, about our place in the community, about did I spank my children,
> did I do this or that—it was *just awful*. And I started crying. I just broke
> down. And I said, "I cannot *bear* this any longer—the pressure of this.
> What are they wanting to *know* about me? What are they *wanting* to
> know? All *I* want is a plan of remediation. That's *all* I want."

In his explanation of the disciplining processes of power, Foucault (1978) shows how confession is used in combination with examination in the form of "the interrogation, the exacting questionnaire . . . reinscribing the procedure of confession in a field of scientifically acceptable observations" (p. 65). Indeed, this scenario suggests imagery of a police interrogation, a discipline that routinely relies upon confession. In her position as "the suspect," Mimi must answer rapid fire questions posed by "the authorities," questions she recalls as being "the most incredible *penetrating* questions you have *ever* heard in your life [**voice lowers**]" (lines 1012–1013). As she is not made privy to the knowledge the psychologist holds, Mimi does not know how she "should" respond nor does she understand how her responses might be interpreted or for what purpose—reflecting a significant power differential between professional and mother. Mimi concludes her memory of this confessional scene with a description of how she "just broke down" (lines 1016–1017) under the pressure of such interrogation.

In the narrative that follows, Cam recalls an incident in which a professional uses particular language to describe not only her child, but also *herself*.

C: I remember we had her tested by one man here in
155 Pleasantville . . . this was probably like in 1976, I guess. But I remember
I was getting ready to go in the hospital and have a hysterectomy. And I
was just *really* down. It was the *day* before I was to be admitted to the
hospital. I was just *really* down because I physically had to have the
hysterectomy, but I *knew* I couldn't then have any more children and it
160 was emotional for me. So, Bob [her husband] said, "I'm going to go to this
conference to get the results on this testing on Pat." And he came home
and he wouldn't tell me what the guy said. Later *after* surgery, when we
sat down and talked about it—he said, "Well, he said the *only* problem
that Pat had was that she was *retarded* and that she had a mother that
165 wouldn't accept it. And if her mother would *ever* accept that she was
retarded, then ya'll could go from there." Those were all slaps in the face,
you know, as if to say there's something wrong with *me*—I can't see what
I'm seeing.

In this recollection, the professional fingers Cam as an active contributor to Pat's pathology because of what he identifies as *her* pathology, namely, the refusal to accept the reality he terms mental retardation. Retelling this story from the vantage point of the present, Cam openly rejects such authoritative discourse, reclaiming her dismissed knowledge by stating that it was "as if to say there's something wrong with *me*—I can't see what I'm seeing" (lines 167–168), leaving unsaid the implication that there is, instead, something wrong with the professional. She describes the experience as "all slaps in the face" (line 167), imagery that again reflects dominion over her. Her narrative is made all the more poignant by the juxtaposition of her loss of *corporeal* capacity for motherhood alongside the professional doubt cast upon her *psychological* capacity for motherhood. It is worth noting that

Bob, the father, is not implicated by the professional as either possessing pathology himself or contributing in any way to Pat's pathology.

Despite her intense discomfort with both the evaluation process and her interactions with professionals, Cam continues to seek out professionals who define her daughter in language she resists, illustrating how we are always already engaging with the discourses available to us. Haug et al. (1987) describe such subjectification as "the process by which individuals work themselves into social structures they themselves do not consciously determine, but to which they subordinate themselves" (p. 59). Cam recollects her experience of yet another private evaluation during the 1970s, this time moving within the hierarchy of authoritative knowledge to a "higher level" expert at an out-of-state university.

C: When she was in the third grade, so she was probably eight or nine, I took her to the University of Tennessee at Memphis and spent a week in their Child Development Center being tested. Pat and Bob and I flew up on Sunday. Bob and I met with them on Monday morning and then he flew home. Pat and I stayed for the week and then he flew back on Friday.
180 Basically, on Friday after a week of testing, they told me that our child was *retarded*, that she needed to be *institutionalized*, that we needed to go on with our lives, to put her somewhere and let her be, and we should go on with our lives and raise our other child. I remember flying home that night. Pat, Bob, and I were flying home from Memphis that night and
185 I was *completely devastated*. And we were flying through a thunderstorm. You could see the lightning out the window of the plane. I remember thinking, and I'm not a person who has negative thoughts, but I remember thinking, "If this plane goes down, it's all right because I *cannot do* what they're telling me to do. If that's the way I have to live my life, I *simply*
190 cannot do it." It was one of those times where it's like I don't know if I can go on with life. This *pain* is so *great* because *I* know this child has ability, but *nobody* else can see it. And all the professionals keep telling me, one after the other, that she's retarded, that she can't do anything.

In Cam's recollection, the experts not only regard Pat's body as a discursive site upon which to inscribe notions of abnormality (mental retardation), they assume authority to determine where Pat's "abnormal body" should be placed in society (an institution), a space that presumably serves the best interests of everyone. Inherent within the expert language is a "discourse of disability" in which it is deemed natural and right to subjugate individuals classified as "abnormal" to the lowest status possible—absolute exclusion from participation in society.

Despite emergence of a competing "discourse of normalization" (Wolfensberger & Nirje, 1972; Blatt & Kaplan, 1966) in the 1960s and 1970s, institutionalization remained the predominant socially constructed consensus for treatment of mental retardation well into the 1970s, as is evident within Cam's narrative. Implicit within the advice to "go on with our lives, to put her somewhere and let her be" (line 183) is a professional ideology of disability that classifies Pat as less human than the other members of her family who "deserve" to live their lives disencumbered from her. On the other hand, Pat should, on the basis of her disability, live her life secluded from her family and society. The experts further dehumanize Pat in their recommendation that her parents focus on raising their "*other* child" (line 184), who by virtue of his normality is deemed worthy of parental love and attention.

Despite being "completely devastated" (line 185) by the language used by experts to describe her child, Cam refuses to engage at any meaningful level with a discourse that neither fits with her own knowledge of Pat (confirmed by a subsequent diagnosis of language impairment/learning disability) nor her understanding of what it means to be a mother. Thus, Cam's resistance to this powerful discourse illustrates Foucault's (1980) assertion that although "power *is* always already there, that one is never 'outside' it" (p. 141, emphasis in original), it does not mean that "one is trapped and condemned to defeat no matter what" (p. 142). Cam's thoughts during the harrowing plane ride home speak to the depth of emotional pain sustained from this experience: "'If this plane goes down, it's all right because I *cannot do* what they're telling me to do. If that's the way I have to live my life . . . I don't know if I can go on with life'" (lines 188–191). In the face of authoritative proclamations, Cam refuses to abandon what *she* knows to be true about her own child. It is noteworthy that each of the five mothers reports incidences of wide discrepancy in the language used by different experts to explain their children, lending credence to the socially constructed nature of disability (Linton, 1998; Corker & Shakespeare, 2002). Referring to the fact that experts label her child as both "normal" and "abnormal," Della challenges the validity of professional knowledge: "So, in trying to do something for her, how do you know what to do when you get all these different opinions? I think it goes back to—*nobody* knew." Cosette relates a similar experience of receiving widely discrepant professional explanations regarding her eldest child, Katie.

170 C: When she was five years old, I remember when she was
tested to get into preschool and the fellow that tested her said that he
had never seen a child at that chronological age score as high as she
did on that particular exam . . . She went into school and her teachers began
175 telling me that she was not completing assignments on time and that
she seemed to be easily distracted. She was very, very difficult at home,
a very irritable child, basically noncompliant, a poor sleeper, poor

relationships with her siblings. And then when she was in the third grade,
I did take her to be tested at the request of the pediatrician. We went to
180 a local psychologist who did the battery of tests on her—the Wechsler,
I recall I think. And after the exam was over, she looked at me across
the desk, very coldly, and said, "Your daughter has brain damage." So,
that was quite a shock. I knew that I had a difficult, unhappy child, but
I certainly wasn't prepared for that.

Cosette, having had a professional describe her child's abilities at five years of age as surpassing all other five-year-olds that he had ever tested, is left to cope, just three years later, with a professional who labels her child as brain damaged. It is of additional interest that the second evaluation takes place shortly after the implementation of PL 94–142. Despite the merging of medical, educational, and legal discourses within the newly established system of special education, Cosette, a registered nurse, engages instead with the medical community. She looks to her pediatrician for help with her child's learning and behavioral problems who, in turn, refers her to a private practice psychologist rather than to the school system. The psychologist, operating outside the discursive practices of the newly established special education system, defines Katie's learning and behavioral problems in terms of "brain damage" (line 182)—language already replaced in favor of "learning disability" by socially constructed consensus among special education professionals. Thus, Cosette presents her child for evaluation during a time in which competing discourses lay claim to "the reality" of children who display problems in learning and behavior. Thus, at this particular historical moment, Katie becomes constituted as *being* brain damaged by a discourse steadily losing validity to discursive practices that construct children with learning problems as *being* "learning disabled." In other words, it would not have been possible for Katie to be constructed as brain damaged within the public school system at this time because there was no language available to describe her in this way.

Cosette goes on to describe her emotional response to the language used by this professional to describe Katie.

190 C: I remember it very clearly. It was a relatively small office. She had
asked my daughter to go outside in the hallway for a few minutes.
And, she just looked at me as if you look at someone and say, "How
are you doing today?" and she said to me, "Your daughter has brain
damage." And I felt at that moment that I had just been *shot*, I guess.
195 The first thoughts that came to mind were how we apparently had
been abusing that child, if indeed she did have brain damage, that
we were expecting her to do things that she was not capable of
doing and expecting her behavior to be a certain way if she truly
was not capable of doing that because she had brain damage. It was
200 a very awful feeling [**voice lowers**].

Cosette signals the significance of this memory when she states, "I remember it very clearly" (line 190). Within this scene, Cosette contrasts what she believes to be the psychologist's experience of this interaction with her own. She recalls the cool, detached manner in which the psychologist delivers the information in stark comparison with her visceral reaction upon receiving it. Cosette likens the experience to a random act of physical violence that materializes out of nowhere ("I felt at that moment that I had just been *shot*" [line 194]), irrevocably altering life as she knows it. In a flash, Cosette doubts her own understanding of Katie as well as the ways in which she has mothered her so far, a material consequence that will plague her for years.

During the latter half of the 1970s, special education discourse, emerging out of PL 94–142, becomes *the* socially constructed consensus to define the way that public schools respond to children with disabilities. Subsequently, special education becomes institutionalized within public schools, giving rise to an apparatus through which the new discursive practices can circulate. At this historical moment, the discursive practices of medicine and law merge with education, creating new ways to speak about and act upon children with disabilities. As could be anticipated, however, the introduction of such a system (and its accompanying ideology about disability) into the existing institution of public education does not take place without resistance. Mimi, recollecting the day in 1978 when her son attended his first special education class at the local elementary school, illustrates this point.

M: The principal of the school was a man who was quite sloppy. He had a Ph.D. or something, but anyway, he was very sloppy. He didn't run the school in a very tidy manner—just very laid
640 back. One day when I was up there, he complained. It was the day actually that I went up to take Pete for his first day. And he *complained* to me that he had the Southeastern Center of Pleasantville County in his backyard [**slight pause**]. And I said, "What do you mean, Dr. Connor?" And he says, "*Look* at all these portables! I have *all* of these
645 portables here. This is for *retarded* children. This is for *LD* children. This is for so and so, this is for so and so." And I said, "Dr. Connor! Do you know where Pete went to school before he came here? And I am so thankful for it?" He said, "Where?" I said, "The Southeastern Center. I can't *believe* you have this attitude toward these children!"
650 "It's just a *pain* to me!" he said. I mean, he didn't back up *at all*! "It's a *pain* to me [**strong emphasis**]!!"

Mimi, whose son attended the Southeastern Center (a self-contained public school for children with disabilities in a neighboring county) prior to the implementation of PL 94–142, remembers the joy of exercising her legal right to have Pete, now in fifth grade, attend the local public school. However, Mimi's enthusiasm about her son's newly won rights to a public education dampens considerably when she hears the principal openly denigrate, without compunction, the integration of children with disabilities into his school. Dr. Connor's comments, upon surveying the newly arrived portables in the school's backyard, reveal both his resistance to the idea of children with disabilities belonging anywhere other than a separate facility and his perception of having had a considerable, and seemingly unreasonable, burden thrust upon him by the new law. Mimi is stunned not only by the language he uses to talk about children with disabilities, but also that he does so in her presence as the mother of a child with a disability. Incredulous that the principal does not "back up *at all*" (line 650) once she confronts him, Mimi has an inkling, on this very first day of school, that her child may have won the right to *be* in public school, but that the promise of a free and appropriate public education may remain an elusive reality.

As the apparatus of special education moves into place within public schools during the latter half of the 1970s, so does the agreed upon classification strategy for determining who is eligible to receive the newly established services. Thus, the testing industry, along with its professional workforce, takes center stage as the sanctioned mechanism for sorting children into normal/abnormal. With the institutionalization of testing within public schools, parents possess the legal right to request, at no cost, psycho-educational assessment of their children. Lindsey recalls the first time she exercised her newly established legal right to have her son Charlie evaluated by the public schools.

L: 60 And I put him in a *pre-first* class that had 9 little boys in it
and thought he just needed another year to grow. And he did
okay. He was reading some by the end of the year, but he was *not*
at the top of that class. And as bright as he was, I was a little bit
clueless as to *why*. And at the same time, I was starting my masters
65 at the university there and just talking to some of the people there.
I asked the public school system to test Charlie before he entered the
first grade. And so, during the summer before first grade, we went
out to a child development center. It was really funny when I think
back on it. They had a basement room with a picture glass window
70 and they set Charlie in front of the window. I didn't know this until
I went back for the assessment review. But they had tested him sitting in
front of this window which opened onto a playground [**ironic laughter**]!
This is where he is supposed to get his IQ and I *knew* that, you know, there
were some *attention* difficulties [**laugh**], too. *But* I was very surprised
75 when they told me that he had an IQ of—I believe it was—91. And they
said that, you know, he was just the *sweetest* little boy, had the nicest

> blue eyes, but that he was just really *not that smart.* And that going into first grade that they would tell the teacher and they would put him in the bottom reading group and they just wouldn't pressure him and that he
> 80 could move at his own pace. And he would be starting school at almost seven and a half. So [**sigh**], you know, I tried to balance between accepting my child for who he was and not really *believing* that this could be *right* [**tone of incredulity**].

Lindsey's narrative illustrates how "the language of testing" begins to operate as a tool that possesses the authority not only to *name*, but also to limit. In this situational context, Lindsey, a graduate student in early childhood education, is unable to reconcile why her bright young son Charlie struggles to acquire basic reading skills. Seeking help to better understand how to facilitate Charlie's learning, Lindsey eagerly looks to the newly implemented system for the evaluation now provided under PL 94–142.

After the evaluation is completed, Lindsey returns for the feedback conference. In describing this incident, it is of interest that Lindsey refers to the persons conducting the meeting only as "they." She mentions no other identifying features of these persons, instead relating her recollection through a disembodied ventriloquism of the language that "they" used in this meeting. It is also of note that she portrays only the voice of the anonymous evaluators, representing herself as having no voice in the meeting. This is in striking contrast to lines 60–74 in which she presents herself, from the vantage point of the present, as a knowledgeable agent capable of critiquing the professional knowledge she is about to describe. For example, she expresses little faith in the judgment of professionals who would test her son in front of a window that opens onto a playground, punctuating her low appraisal with an ironic laugh. She goes on to say, "This is where he is supposed to get his IQ and I *knew* that, you know, there were some *attention* difficulties [**laugh**], too" (lines 73–74), suggesting the superiority of her own knowledge of her child. Moreover, in describing the evaluation as a place where he is supposed to get his IQ, Lindsey's wording reveals her sense that Charlie's intelligence will be assigned to him at the evaluation site, constructed in the act of professional assessment, as if his intelligence, undocumented, did not exist before.

Lindsey's manner of recollection regarding the feedback conference suggests her perception of being positioned at the time not as a co-contributor as the law requires, but a mute recipient of authoritative knowledge. In line 74, Lindsey begins her sentence with an emphasized "but," signaling that what is to follow is her actual belief or response, which is that she was "very surprised" at being told that Charlie had a measured IQ of 91. As she goes on to relate what "they" tell her next, Lindsey mimics their paternalistic rhetoric, "And they said that, you know, he was just the *sweetest* little boy, had the nicest blue eyes, but that he was just really *not that smart*" (lines 75–77).

Her ironic tone of voice belies her unmistakable feeling that "they" believe that Charlie will not and cannot amount to much given his documented lack of cognitive potential. Lindsey gives evidence for this feeling as she reports that "they" go on to suggest that Charlie be placed in "the bottom reading group and they just wouldn't pressure him and he could move at his own pace" (lines 78–79)—all euphemisms to support their belief that he was "just really *not that smart*" (line 77). Thus, Lindsey begins to sense what Foucault (1975) refers to as the exercise of "a power that insidiously objectifies those on whom it is applied" (p. 220). Despite the fact that her own knowledge of Charlie, from her vantage point as mother and early childhood educator, does not fit with this professional assessment *and* that she questions the validity of the assessment after becoming aware that Charlie was tested in front of a window that opened onto a playground, Lindsey yields to the power imbued within professional judgment to determine "who he was" (line 81), thereby limiting her child's possibilities: "So [**sigh**], you know, I tried to balance between accepting my child for who he was and not really *believing* that this could be *right* [**tone of incredulity**]" (lines 81–83).

In the following narrative, Mimi remembers a similar incident in which a professional presents a different and conflicting reality that assumes superiority, over her own, on the basis of science. Mimi recalls taking her then 14-year-old son, identified since first grade as language impaired and LD, to be evaluated by an internationally recognized learning disability expert at an elite university. Mimi, like Cam, believes that "the answer" must reside at the apex of the professional knowledge hierarchy—the research university. Indeed, Mimi's memory of her family's preparation for the trip is reminiscent of an archetypal plot in which the heroine sets out on a quest to find the source of pure and perfect knowledge.

M: And so, we prepared to go. Bought a new motor home. We had a Class C motor home and we bought a brand new motor home—a Class A motor home. We had never even been out in it as a matter of fact. It had a
905 big wide front seat put in especially so Pete and Sonny [her youngest son] and I could all sit in the front seat. Sonny at that time was six years old. We all got in the motor home and headed [there].

And not unlike the heroine of all quest narratives, Mimi is to learn that the source of true knowledge emerges not from a guru (expert) atop some mythical mountain (research university), but from within herself in the process of her journey. She describes her visit as a dire disappointment.

917 M: So, actually it gave us *nothing*. It really gave us nothing. There was nothing. But we came back and we started high school and *you* became his teacher.

Mimi reaches the point in her narrative in which I enter as a character in the unfolding plot. Unlike characters who cannot speak for themselves within a narrative telling, I, in my dual role as character *and* audience, am able to participate in the construction of Mimi's narrative by sharing my recollection of the same event. I interject, at this point, because I am intrigued that Mimi relates this experience with flat affect, repeating in three consecutive sentences that nothing came out of the evaluation. I, on the other hand, clearly recall that the evaluation had had a traumatic effect upon her. I gently suggest what I remember this professional saying to Betty, her graduate student at the time and a former colleague of mine, after evaluating Pete.

920 J: It seems to me that what I remember . . . about that story was that she said to Betty, "Why didn't you *tell* me that Pete was mentally retarded?"
M: There's something about that coming back. And there was something
925 actually in the papers, I believe, yes, I remember that now.

Having had her memory jogged by my own, Mimi appears somewhat stunned by both the reintroduction of this information into her consciousness and the realization that she had blocked it out in the telling of her story as if it had never happened.

M: That was one of those things that I have put back there because it was so painful, *but* I now remember—I can't remember whether I learned that while he was at [the university] or whether it was when her report came.
930 We'll have to look at it and see. It's right up here. Anyway, I remember then that I went—I was on the school board then—and I went up to Patti Smith's office and went in and sat down and cried and cried and cried and cried and said, "Is he? Is he? Why haven't you told me?" She kept saying, "He isn't! He isn't! You have to understand that he isn't." The *pain* of
935 that, interestingly enough, I can't remember much more about it.

Once she recalls the results of the evaluation, Mimi immediately resurrects the depth of her emotional response, describing how she sought out a trusted special education administrator who had been involved in Pete's education for years and "cried and cried and cried and cried" (lines 932–933), certain that she had been operating under the erroneous assumption that her son is *not* mentally retarded while the experts have known otherwise. The memory of this event is so intense that Mimi concedes, "The *pain* of that, interestingly enough, I can't remember much more about it" (lines 934–935).

Conflicts in Shared Decision Making and IEP Implementation

During this historical era, PL 94–142 emerges to grant educational rights not only to children with disabilities, but also to their parents. The law guarantees parents the right to collaborate with professionals in making educational decisions regarding their children with disabilities. Specifically, the law ensures that parents have the right to be informed, the right to be knowledgeable about the actions to be taken, the right to participate, the right to challenge, and the right to appeal (IDEA, 1990, 1997, 2004). As such, the institutionalization of parental rights under PL 94–142 signals the introduction of a set of what Foucault (1972) calls "statements"—those presuppositions circulating as true and accepted in a given society—and out of which structures arise to determine *who* has the right to make statements, from *what sites* these statements originate, and *what* positions the subjects of discourse inhabit (Dreyfus & Rabinow, 1983).

Prior to the passage of PL 94–142, exclusion reigned as the socially agreed upon response to children with disabilities. The practice of exclusion rendered invisible the child with a disability—whether warehoused literally out of sight in a separate educational facility or simply ignored in a general education classroom as being beyond anyone's responsibility. Della recalls the invisibility of her daughter during the 1960s.

D: I remember talking with the principal when she was in the third
grade that I was concerned [**slight pause**]. And he didn't know a *thing*
145 about Jennifer. He didn't know a *thing* that could be done [**slight pause**].
He showed *no* interest [**pause, voice lowers**] *whatsoever* . . . I really think
that he thought I was trying to tell *him* how to run his business [**light**
155 **laughter**] . . . [**pause**] Back then, the classes had 30 and over kids in
them, so I understand that the teacher did not have time to give to one
child like Jennifer needed. But, I don't think her third grade teacher [**slight**
pause] who was an experienced teacher and was *thought* to be a *very* good
teacher—she [**pause**] didn't show [**speech slows**] much concern for
160 Jennifer or [**long pause**]—her inability to learn [**voice lowers**].

The extent of Jennifer's invisibility is well illustrated in Della's recollection of her meeting with the principal. Not only does the principal have no knowledge of her child, Della also recognizes that he has "no interest *whatsoever*" (line 146) in engaging with her at any level regarding Jennifer's education. In daring to draw the principal's attention to her daughter's educational plight, Della senses she has transgressed the unstated consensus for "how things are" for children who do not learn: "I really think he thought I was trying to tell *him* how to run his business" (lines 152–153), and indeed, he responds by opening no space for her contributions. Initially, Della seems accepting of the presupposition that teachers could not and should not be expected to give the kind of attention "that a child like Jennifer needed" (line 157). However, she prefaces her next statement with "but" (line 157), signaling that she is about to reveal what she really thinks, and goes on to expose Jennifer's third grade teacher, whom others "*thought* to be a *very* good teacher" (lines 158–159), as someone with complete disregard for her child's learning difficulties. In making this statement, Della pauses between phrases, slows the rate of her speech, and lowers her voice, as if still hesitant, all these many years later, to rightfully name the injustice of her child's education.

By the early 1970s, parents and advocates across the country begin to draw upon civil rights discourse to challenge the presupposition that it is natural and right to exclude children with disabilities from public education. Mimi recounts one such meeting that took place in 1974 between school district personnel and parents whose children had been labeled LD by medical or psychiatric professionals.

M: They were just having a meeting of Pleasantville County Parents of
LD Children. This is a parent group. They were wanting to talk to people
in the school district. The school district people were actually speakers
420 that night—question/answerers who got confronted with *big* questions.
I sat and listened and realized that we had a lot of *angry* parents in the
county. And that I was *one* of them, but I didn't say anything. I just kept
my mouth shut. When the meeting was over, then *I* went up to these
school district people and said, "I am Mimi Bing and I have come here
425 because I can't get you to return my phone calls. I saw that you were
going to be here tonight and I am *here* and I want you to talk to me about
my child." I can't remember if it was at *that* time or later that I told them
that I had found the place I wanted him to go. And that I wanted them to
pay for it. That took a lot of courage . . . The next thing we knew—after we
got a lawyer and we did actually file suit against the school district—it
465 never took place because they signed a multidistrict agreement and he
went to school at the Southeastern Center the very first year!

Mimi's recollection of this forum exemplifies the momentum of resistance created by parents openly challenging exclusionary practices. She notes that "the school district people were actually speakers that night" (lines 419–420), implying that, as speakers, they positioned themselves to control the meeting; however, they quickly become positioned by parents as "question/answerers who got confronted with *big* questions" (line 420). Sitting and listening to other angry parents, Mimi realizes that she is not alone, but rather "*one* of them" (line 423), part of a growing contingency. Mimi's narrative is a useful illustration of Foucault's (1980) assertion that

> in reality, power in its exercise goes much further, passes much finer channels, and is much more ambiguous, since each individual has at his disposal a certain power, and for that very reason can also act as the vehicle for transmitting a wider power. (p. 72)

After an exhaustive search for an appropriate placement for her young son, Mimi finally locates a public special education school in a neighboring county. The school agrees to take Pete, but requires an out-of-county tuition fee. Bolstered by the momentum of resistance among parents, Mimi reasons that it is the school district's responsibility to pay the tuition because they have no class for Pete to attend. Reflecting back upon the experience, Mimi acknowledges that it "took a lot of courage" (line 429) to file such a lawsuit *before* the existence of PL 94–142. Perhaps sensing winds of change on the horizon (and/or choosing not to antagonize the Bings who own a small-town newspaper within the county), the school district signs a multidistrict agreement and pays Pete's tuition for the next four years—until required under PL 94–142 to provide their own services for children with disabilities.

As the requirements of PL 94–142 move into place within public schools, parents of children with disabilities begin to exercise their newly won right to collaborate with school professionals about educational decisions. In fact, the new law guarantees parents of children with disabilities a particular set of rights within public schooling never before afforded to parents and not available to parents of children *without* disabilities. The law's requirement of collaboration between professionals and parents offers a vision for a new era of relationships—one in which the long-standing tradition of professional dominance within public schools might give way to authentic alliances with parents. However, as PL 94–142 becomes institutionalized within public schools, scientific, legal, and educational discourses intertwine to form a new and hybrid discursive practice. Rather than a means to disrupt traditional patterns of professional dominance over parents, the new law functions in a way that appears to *reinforce* professional knowledge—bolstered now by its triumvirate grounding in science, law, and education—as knowledge exponentially superior to that which a parent brings about his or her own child. Cam illustrates this point in the following recollection about her meetings with school professionals in the late 1970s.

C: So, I had to get strong enough
720 within my*self* to be able to walk into the meeting, and regardless of what
they said, I knew where I stood. And what they said did not get me. I used
to say to myself, "It's water off a duck's back. It's water off a duck's
back." It was just words, but I would sit in a meeting and think when they
were saying things that I didn't want to penetrate my brain, I would just
725 say "It's water off a duck's back. It's water off a duck's back" to myself—
so that I could *cope* and come back and say what *I* wanted to say. Because
what would happen to me as I'd get in this meeting with these people and
they would start telling me what the tests showed and all that kind of stuff,

> they would rattle my cage. And so, my brain would start going like
> 730 "Oohh!! What am I supposed to say now? What am I supposed to do now? How do I counteract that? I know she does this, but what do I say back to that? I don't understand that test [**rapid speech to reflect the pressure**]!" So, I had to get strong enough within myself that I could let them say *whatever* they needed to say and I was going to say what *I* knew to be
> 735 *truth*. And I wasn't going to let *them* shake my truth in what I knew about Pat. And that took a *process* of time because I had nobody to really counsel me and help me make that transition. Honestly, it was a very difficult place to grow to. It was a *painful* place. It was a psychologically challenging place to be. But it was a necessary place. So, I had *arrived*
> 740 there and I was like a mama bear ready to fight for her cub. It didn't matter about *me* anymore. Nobody was going to make *me* feel guilty, nobody was going to shake *my* confidence, nobody was going to tell me I was *wrong*. I was going to fight for what was best for Pat.

Although legally granted *access* into the arena of educational decision making, Cam recalls a less than satisfying experience of collaboration under the new law. It is unclear exactly with whom Cam engages in these meetings as she only refers to "they" (lines 721, 723, 728, 729), "these people" (line 727), and "them" (line 733)—reflecting her sense of alienation from the school professionals involved. Cam's consistent use of third-person plurals (in contrast to using first-person plurals such as "we" and "us") signifies her perception of being positioned in a passive role by an authoritative "them." It is of further interest that Cam describes engaging with professional discourse in terms of "what would happen to me" (line 727). Indeed, Cam's verbal expression is rife with imagery that suggests resisting assault. For example, Cam opens her narrative by describing how she had to "get strong enough within my*self* to walk into a meeting, and regardless of what they said, I knew where I stood" (lines 719–721). Thus, Cam signals that the story she is about to tell includes a conflict of significant proportions—one requiring a warrior's mantle of strength and resolve. It is noteworthy that Cam marks her entrance into the meeting by using verbs (e.g., walk, stood) that reflect agency. Once the meeting is under way, however, Cam becomes positioned in a role that requires her to sit and listen. She recalls inventing a strategy that enables her to "*cope* and come back with what *I* wanted to say" (line 726), clarifying that "when they were saying things that I didn't want to penetrate my brain, I would just say, 'It's water off a duck's back. It's water off a duck's back' to myself" (lines 723–725). Thus, Cam experiences her engagement with professional discourse as an exercise in deflection—accomplished by visualizing herself repelling the barrage of words in the same way that a duck's back repels water.

Assault imagery appears again in Cam's recollection of the way in which school professionals present her daughter's test results. Rather than being asked to engage in collaborative conversation about her daughter's education, Cam recalls immediately being positioned as a mute recipient of a rather aggressive professional discourse. For example, she recounts that "they would just start telling me what the tests showed and all that kind of stuff and they would rattle my cage" (lines 728–729). Cam takes us inside the mind of a mother as she listens to the professional language used to describe her daughter. As she describes what she is thinking while the professionals talk, Cam speaks in a rapid and pressured manner to reflect her intense anxiety in trying to comprehend *and* respond to an unfamiliar discourse: "Oohh!! What I am supposed to say now? What am I supposed to do now? How do I counteract that? I know she does this, but what do I say back to that? I don't understand that test!" (lines 730–733). Cam sees school professionals operating out of an advantageous position—controlling what can be said, how it should be said, and who can say it—effectively situating her to accept or react to what *can only* be said in contrast to opening a space for her to contribute to what *could* be said.

In contrast to her earlier narrative in which she describes "the language of blame" used by professionals, this narrative account, taking place a few years later, reveals Cam's growing sense of agency in response to the powerful professional discourse that alone defines her daughter. It is noteworthy, for example, that Cam uses "I" 26 times within 24 lines of text—evidence of her increasingly assertive stance. Cam's resistance is unmistakable in the following excerpt: "So, I had to get strong enough within myself that I could let them say *whatever* they needed to say and I was going to say what I knew to be *truth*. And I wasn't going to let *them* shake my truth in what I knew about Pat" (lines 733–736). Thus, Cam repositions herself in relationship to professional discourse, conceptualizing herself as possessing the power to *let* them speak as well as the power to assert her right, as a mother, to present her own truth. Cam's turning of the tables is evident in her dismissal of professional knowledge as "*whatever* they needed to say" (lines 733–734), contrasting with her own knowledge framed as "what I knew to be *truth*" (lines 734–735).

Reflecting upon her transition from passivity to agency, Cam likens the experience to a long, arduous, and solitary journey. Metaphorically speaking, Cam finds her power in this "necessary place" (line 739), returning "like a mama bear ready to fight for her cub" (line 740)—imagery that suggests the anticipation of conflict, rather than collaboration, in her interactions with school professionals. Cam's concluding statements—"Nobody was going to make *me* feel guilty, nobody was going to shake *my* confidence, nobody was going to tell me I was *wrong*" (lines 741–743)—indicate a defining moment, a point after which Cam refuses to be positioned as passive ever again. Shortly after the introduction of PL 94–142 into public schools, Mimi and her husband attend an eligibility and placement meeting as required by the new law.

570 M: They called us to a meeting at the school district office. There were at *least* 12 or 13 school district officials there. There was, of course, the social worker. There were all these people in the hierarchy—all the way up to everybody but the superintendent was there. They started off by saying, "Mrs. Bing, would you tell us about your child?
575 Let's just start at the beginning and tell us about your child." Well, *thinking*—and *every* time you go to one of these—this is what you think—you think, "I *have* to do this because *this* is what's going to help my *child*." So, I began at the beginning like I did with you just now. I began at the beginning. Of course, there were many more details that I've
580 forgotten now that I told. And for about two *hours*, I *told* them the story of my child's life. Cast my pearls before swine is what I did. I sat there and I *cried*. And they watched me cry. I went through the *painful* parts—I mean, there's just so much *pain* in it—and still we were in the *midst* of pain because we were trying to decide on next year's education plan.

Mimi begins by describing the scene into which she and her husband enter—a room in which they face "at least 12 to 13 school district officials" (line 571), "the hierarchy—all the way up to everybody but the superintendent" (lines 572–573)—illustrating the power differential inherent in the number of school district officials present and the status each represents. Given her prior interactions with the school district regarding tuition payment, Mimi is aware that school officials regard her as problematic, a perception underscored by the gathering of troops at this placement meeting. When asked to tell about her child, however, Mimi puts aside her sense of mistrust and pours out her heartfelt story, believing that, in doing so, her child might benefit from enhanced understanding on the part of school officials. She states ruefully, "Cast my pearls before swine is what I did" (line 581), referencing a biblical adaptation from Jesus' Sermon on the Mount that means to share something of value with people incapable of appreciating it (Hirsch, Kett & Trefil, 2002). Mimi exposes deeply personal and painful aspects of her child's life, expecting compassion and understanding from her audience. Instead, she recalls that "they watched me cry" (line 582), suggesting a dispassionate and somewhat voyeuristic response to her story. Mimi's willingness to share intimate details within this unlikely context perhaps reflects Foucault's (1978) notion that "the obligation to confess is now relayed through so many different points, is so deeply ingrained in us, that we no longer perceive it as the effect of a power that constrains us" (p. 60).

Maintaining a stance of objectivity, the committee turns abruptly to the business of the meeting.

M: We
585 finished and they went around the room and they *all* had a little chat—something that they said—not particularly to *me* but to each other about the plans for Pete and all that. I kept *waiting* to be presented with a teacher's name, at *least*, and this place they were going to put him and there was *nothing*. Finally, this person who was conducting the meeting
590 said, "Mr. and Mrs. Bing, we want to tell you now that we don't have a place for Pete right now. But when we do, we will be in touch with you. We're working on this teacher we are thinking about hiring." And I said, "Do you *mean* that *you* brought us here and I've *told* you all this and this is *not* a placement meeting and this paper says this is a placement
595 meeting? And this is *not* a placement meeting? And it's *not* a staffing meeting? This is *nothing*?! What *was* this?" They said—sort of like—"It's okay. Don't worry. We're going to get you a teacher [**flippant tone**]." On the way home, I became *very, very, very* angry . . . I said to my husband, "They are *never* going to do this again. This is *never* going to happen again . . . This is not going to happen again." I came home
605 and I wrote a letter to the superintendent. I told him about what had happened and about how I had shared with them all this stuff and that they had called us over there. I copied the form that we had to sign to go into the meeting. And the meeting was obviously *not* that. Then I copied a portion of the law that said that at *every* meeting, there had to
610 be a person at the meeting that can make the *final* decision. So, I placed that before him in the letter and I said, "Do not *ever* again call us to a meeting—if it *means* that *you* have to be *there*, Dr. Brown—then we suggest that you come because we are not coming to any more meetings that waste our time and so forth." So, we didn't go to any more
615 meetings like that [**light laugh**]! That was just the *beginning* of the battle—just the *beginning* of the battle.

Despite the guarantee of parental rights under PL 94–142, Mimi's recollection suggests that these school professionals, working in the late 1970s, continue to operate out of a framework that positions parents as passive. Although invited as a participant to the meeting, Mimi recalls that the school professionals "went around the room and they *all* had a little chat—something that they said—not particularly to *me* but to each other about the plans for Pete and all that" (lines 585–587), leaving her on the periphery to patiently await their conclusions. Thus, Mimi discovers that the new law may guarantee her presence at the meeting, but it does not necessarily guarantee shifts in professional attitude.

The new law does, however, provide Mimi with legal grounding for her outrage upon learning that the school district has neither a classroom nor a teacher for Pete. In response to what she perceives as a lackadaisical attitude toward the process, Mimi counters with legal documentation, initiated from *within* the school district, of her invitation to a placement meeting. With documentation in hand, she challenges their level of responsibility toward her and her child by asking, "And this is *not* a placement meeting and this paper says this is a placement meeting? And this is *not* a placement meeting? And it's *not* a staffing meeting? This is *nothing*?! What *was* this?" (lines 593–596). Much like Cam's defining moment narrative, Mimi's experience represents a turning point for her. She vows that school officials "are *never* going to do this again" (lines 602–603), restating her resolve in line 603 ("this is *never* going to happen again") and again in line 604 ("this is not going to happen again"). In concluding her narrative with what Labov and Weletzky (1967) define as a coda ("that was just the *beginning* of the battle—just the *beginning* of the battle" [lines 615–616]), Mimi alludes to the conflict that inevitably emerges from insisting upon rightful accountability under the law.

In asserting her right to participate in, rather than receive, educational decisions regarding her child, Cam describes a similar conflict with school professionals.

785 C: Yes, they got very *frustrated* with me. Because I would say, "This is not
right. What you are seeing is *not* right. Pat *does* need help. Pat *can* learn
something." I became almost [**slight pause**] belligerent in her defense, at
times, in those meetings. I don't know if belligerent is *exactly* the right
word, but like "No! You are not right. I am sorry. And no! She is not
790 going to be in that placement. She will not be in regular class. Period.
I don't know *what* you're going to have to do, but you're going to have
to find a way. *My* child does not belong in a regular class where she
cannot learn." So, I became very definitive in what my thoughts were.
They became defensive with me. I know they didn't like me at all.
795 I would continue to call back and call back and call back. And you *feel*
the disapproval. And for somebody who *likes* the approval of people,
that was *very* difficult for me because my personality is to do what
people *expect* me to do and I was suddenly feeling much disapproval.

As a result of exercising her right to participate, Cam reports a professional response not unlike the principal who responds to Della's overtures, *before* PL 94–142, as if she is trying to tell him how to run his business. Resisting professional knowledge that does not fit with what she knows about her daughter, Cam disrupts expectations that these school professionals, working in the late 1970s, hold about how mothers ought to behave. It is of interest that Cam describes herself as "almost belligerent in [Pat's] defense, at times, in those meetings" (lines 787–788), implying that only exaggerated behavior draws attention. Cam's contributions to the meeting elicit frustration, defensiveness, and disapproval from school professionals, as if she were transgressing rules rather than operating within them.

800 C: *Body* language changed, folders were *closed*, and hands put on top of them like "We can't do *anything* with *this* woman." People sat up straight. I would be sitting across the table and people would sit up straight and square their shoulders like "*Who* does this woman *think* she is?" You know? And sort of dismiss me.

On the basis of her experiences, Cam recognizes that the right to participate is no guarantee of a collaborative relationship with school professionals. In this scenario, it seems that school professionals understand the right to participate as the right to be present. Cam's attempt at offering a differing perspective is met with authoritative posturing on the part of professionals—for example, closing folders, sitting up straight, squaring shoulders. The message she receives about the value of her knowledge is unmistakable: "*Who* does this woman *think* she is?" (lines 803–804).

Like Cam, Mimi continues to insist upon her right to participate in educational decision making. The more assertive she becomes, the more authoritative school district personnel become in response. Before long, school district personnel orchestrate who may speak to Mimi, to whom Mimi may speak, and under what conditions.

730 M: They *never* let the teacher meet with me by herself. I knew *immediately* that I was some more problem for them because they would never let me meet with anybody alone. They would always send the head of special education for the entire county or the person for this area—*always* a person from this area—*always* a social worker—*always* some
735 psychologist—*always* there were never less than five people.

Once targeted as "some more problem" (line 731) for the school district, Mimi becomes positioned as an object of surveillance. In her retelling, it is of interest that Mimi repeatedly and emphatically uses absolute and binaric language to describe her experience, reflecting the authoritarian response toward her. Of particular interest in this narrative is the way in which the ambiguous "they" (presumably school district officials at the top of the hierarchy) are *able* to "never let" (line 732) Mimi meet with school district employees without supervision and to "always send" (line 732) supervisors of their choosing. The process that operates to position Mimi in this way becomes clearer if understood in terms of Foucault's (1975) conceptualization of power.

> The power in the hierarchized surveillance of the disciplines is not possessed like a thing, or transferred as a property; it functions like a piece of machinery. And, although, it is true that its pyramidal organization gives it a "head," it is the apparatus as a whole that produces "power" and distributes individuals in this permanent and continuous field. (p. 177)

In other words, the power inherent within the educational apparatus functions because all of the players—located in their respective positions in the hierarchy—agree to participate in its circulation, making possible "the operation of a relational power that sustains itself by its own mechanism" (Foucault, 1975, p. 177).

Despite being under intense surveillance, Mimi persists in exercising her right to participate in educational decision making. When the time comes for Pete's triennial evaluation, Mimi insists upon her right to a third-party private evaluation, paid for by the school district, on the basis of evidence that the school district speech/language pathologists do not possess the training necessary to conduct the language assessment her child requires. In response to the school district's denial of her request for a third-party private evaluation, Mimi hires an attorney. School district surveillance now extends beyond whom and under what conditions she can speak to include conditions about where she is permitted to be.

M: They wouldn't let me come even on the *campus*
860 of Pete's school any more. I couldn't even put my *foot* on the ground
over there. I could pick him up, but I could not get out of the car. The
teacher could not speak to me. *No one* there was allowed to speak to me.
No one there was *ever* allowed to speak to me . . . Of course, it got to be
875 sort of exciting because you watch these people and they get to where
they're just beside themselves. Even when I go pick up Pete, it was like
they had to be sure that *no one* had spoken to me. Because evidently,
they were having to give some kind of accountability over at the *district*
office.

Ironically, Mimi's attempt to exercise her rights as guaranteed under PL 94–142 results in having other rights routinely offered to parents *restricted* in her case—for example, the right to speak to the teacher, the right to step onto school grounds, the right to come into the school building, the right to speak to various other school personnel. Much like the experiences Cam reports, Mimi recalls that school personnel respond to her as if she were transgressing rules rather than operating within them. Her observation that "evidently they were having to give some kind of accountability over at the *district* office" (lines 877–879) bears credence to Foucault's (1975) assertion that disciplinary power "constantly supervises the very individuals who are entrusted with the task of supervising" (p. 177). Moreover, Mimi recognizes that she creates her own power through resistance, confiding to me with a hint of pleasure, "Of course, it got to be sort of exciting because you watch these people and they get to where they're just beside themselves" (lines 875–876).

Following a series of tense negotiations between the school district attorney and the attorney representing the Bing family, both parties agree that a third-party evaluation (paid for by the school district) will be administered not only by staff at a university-based child development center within the state, but also by a nationally recognized expert to be flown in at school district cost. Having legally established her right to a third-party evaluation by qualified professionals (deemed unavailable within the school district), Mimi is confident that an appropriate assessment will yield accurate and useful information upon which to base sound educational decisions regarding her son. Upon her arrival at the university, Mimi discovers that school district personnel have been sent to monitor the third-party evaluation and will sit in the observation room beside her.

M: Well, we started our testing. [**pause**]. The
965 school district sent Peg Greene there, a psychologist. Peg went, Dr. Mason went [**pause**]. *Two* other people went [**lowers voice**]. I don't remember who they are now. They were there *all four days*—having their lodging, their transportation, everything paid—for *one little boy* to be tested . . . They were sitting on the other side of the window—*grading the test*—
970 while Pete was being tested [**pause**]. It's not believable, is it? So, these four people sat behind the window and Pete could see them, though. He got to where he would sit like this [**demonstrates how he sat with his body turned from the window with his hand covering his face**]. They had to know what the questions were to grade it at the same time. They
975 didn't trust Wellford to score it *accurately* to suit them. They were scoring it at the same time, but not in the room with Pete. So Pete got to where he sat like this [**demonstrates**] with his hand on the side of his face, covering up his eyes so he could not see through the mirror. It was *absolutely horrible.*

Mimi opens with an abstract (Labov & Weletzky, 1967) that signals the narrative to follow: "Well, we started our testing" (line 964). This seemingly innocuous introductory sentence, however, represents what is *really* at issue—a struggle over knowledge. Mimi's assertion of her rightful place in the process is reflected in her choice of wording. She clearly sees herself as an integral participant (as guaranteed under PL 94–142), using "we" (rather than "they") and "our" (rather than "their") to describe her foray into the hard-won, third-party evaluation process. In response to Mimi's persistent assertion of her rights under the new law, school district surveillance of mother and child now extends to the third-party evaluators. It is a striking visual image of the "professional gaze" (Foucault, 1977)—professionals gazing upon professionals gazing upon "*one little boy*" (line 968). School district personnel bodily situate themselves between Mimi and the clinic personnel in an effort to monitor and control the flow of information. In the midst of telling her story, Mimi pauses to muse aloud, "It's not believable, is it?" (line 970).

On the other side of the observation window is the object of study, a little boy, who sits at the center of attention for five adults (four school district personnel and Mimi) seated behind the observation window as well as for the rotating cast of examiners over the course of four days. School district personnel, under direction to simultaneously score the tests along with the examiners to ensure accuracy, sit close enough to the observation window to see the test protocols, thereby making themselves visible to Pete. In response to being positioned as a passive object of study, Pete turns his body from the window "with his hand on the side of his face, covering up his eyes so he could not see through the mirror" (lines 977–978)—and perhaps, more significantly, so that they could not "see" him. Recalling this poignant act of resistance by her child, Mimi concludes the narrative with emotion suggesting nearness of the events rather than the actual distance of more than two decades: "It was *absolutely horrible*" (lines 978–979).

It is worth mentioning that between our first and second interviews, Mimi retrieved extensive files she had kept over the years—for example, report cards, IEPs, evaluations as well as documentation of written exchanges with school district personnel. Mimi opens the second interview by reading from her copy of the university-based evaluation report, relating with astonishment that the evaluators describe Pete as becoming very upset during testing, asking to see his mother, and then clinging to her—implying, of course, that there is something wrong with Pete rather than something wrong with an evaluation process that places a child under the scrutiny of five *visible* adults behind an observation window for the duration of four consecutive days. Mimi also reads from a letter written by school district personnel to the university-based testing center, which states, "That further a meeting be scheduled immediately involving *only* the testing personnel and the designated *experts* from the Pleasantville Country School District to ensure that the comprehensive evaluation has been done." After reading the excerpt, Mimi looks up and repeats, "*Without us*! Without us. Without us!" (lines 229–230)—as if reliving her incredulity anew at such attempts to control the dissemination of knowledge as a means to exert and maintain power (Foucault, 1975).

The meeting eventually takes place with the Bings present, as required under PL 94–142.

Devalued Knowledge of Mothers

The narratives presented thus far support my earlier assertions that professional dominance has long been the traditional response toward parents. As the narratives illustrate, professionals routinely assume superiority in their knowledge (e.g., scientific, bureaucratic) while dismissing "other ways of knowing" that mothers bring to the discussion.

In the five interviews, I asked each mother to define when and where her story first began. It is noteworthy that each mother identifies "the beginning" as the earliest moments of her *own* recognition of differences, *not* the moment of professional diagnosis. In fact, each mother begins her story long before her child enters formal schooling.

As the mother of two older boys, Mimi recalls that her inkling of Pete's differences occurred shortly after his birth.

M: When they
brought him to me, I immediately saw that something was wrong
10 because when I held him he wouldn't conform to my body. He was
quite rigid.

Lindsey, also the mother of two older boys, remembers her early concern about Charlie walking later than his older siblings.

L: Charlie did not walk until he was 19 months old, which was my first indicator that something might be not okay [**voice lowers**]. The other two boys had walked right around a year . . . the pediatrician, *of course*, was saying, "*Absolutely* nothing's wrong . . . he'll be fine."

Likewise, Della and Cosette identify infancy and toddlerhood as the beginning of their uneasiness.

D: Oh, when she didn't sit alone until she was nine months old. And then she
didn't walk until she was about two and she didn't talk until she was about
10 two.

C: She never slept very much as an infant,
a very restless sleeper—cried an awful lot. She started chewing her
tongue—just at random—when she was probably less than two years
155 old. The pediatricians did not seem concerned about that at all, although
they did observe her closely. She was very delayed in walking. She
was 16 months old before she walked.

It is of interest that both Lindsey and Cosette recollect that pediatricians dismissed their concerns about late walking, late talking, and tongue chewing—symptoms identified years later as retrospective evidence of their children's "neurologically-based learning disability" profiles.

Not only do these mothers recall having concerns during their children's early years, they also reveal powerful skills of observation and interpretation, as illustrated in the following four excerpts (Della, Cam, Mimi, and Cam, respectively).

D: Even when she was little—a baby at home—she would not sit still for you to read to her. Maybe a line or two and then she would wiggle out of your lap and would be gone [**pause**]. It was *awfully* hard to read to her.

C: We first started noticing problems when it was time for her
65 to start communicating. And she just didn't *talk*. She would say single words and she would point to things. She had a wonderful communication system with her facial expressions, and her eyes, and her fingers. But she didn't communicate in sentence structures. She repeated the same sentences—the ones she *learned*. She repeated the same sentences over
70 and over again. She didn't start to put words together on her *own* to express thoughts. And so we started to be concerned.

M: I had been doing a lot of studying in trying to figure out what was wrong with him, myself. One of our doctors had
305 checked a lot of books out of the library, the medical library in Pleasantville, for me to read and I had pretty much decided that he was autistic. I knew that he was not *classic* autism because he made eye contact and his behavior was not—he didn't act out at all—he was very withdrawn. But watching things spin around, liking to play with the
310 vacuum cleaner, and doing his hands like this (**Mimi demonstrates hand flapping**), and echolalia—a lot of echolalia—and an enormous amount of frustration if he was trying to put his GI Joe's clothes and they wouldn't go on and he would have this sort of temper tantrum if they didn't go on right. If things didn't *fit* right, then we did have some acting out. But, it
315 had to do with objects fitting into other objects and so forth.

C: When Pat was four years old, she was able to take a wooden map of the United States and put it together. She could
170 put *all* of the states in the correct place. And I *knew* there had to be some ability there. A four-year-old does not do that. Now, I realize that
her memory is so incredible. I didn't understand then. But I knew that she wasn't *completely* without any ability to do anything if she could put the puzzle together.

Looking at these narratives collectively, it is clear that the mothers are able not only to identify behaviors that appear to be developmentally different from other children at a comparable age (e.g., difficulty sitting still and attending while being read to, using single words rather than talking in sentences, displaying echolalia), but they also demonstrate keen analysis of those behaviors. For example, Cam provides a quite sophisticated diagnostic assessment of what she observed about Pat's early language development, noting that "she didn't communicate in sentence structures. She repeated the same sentences—the ones she *learned*... She didn't start to put words together on her *own* to express thoughts" (lines 67–71). On the other hand, Cam recognized abilities in Pat, such as putting a wooden map of the United States together at age four, that reflected higher cognition than her language skills suggested. Likewise, Mimi explains that although Pete did not meet the criteria for "classic autism" (line 307), she observed that he engaged in autistic-like behaviors such as "watching things spin around, liking to play with the vacuum cleaner, doing his hands like this" (lines 309–310). She also recalled that the antecedent to Pete's temper tantrums frequently "had to do with objects fitting into other objects" (lines 314–315).

Clearly, these mothers possess a wealth of knowledge regarding their children's language, behavior, and general abilities prior to seeking and/or receiving any formal diagnostic information. Yet, their subsequent narratives suggest that professionals not only fail to recognize the rich knowledge they bring, but also dismiss the knowledge that they offer. In the following narrative, Cam describes how professionals in the early 1970s deflect her attempts to share relevant contextual information that conflicts with their assumptions.

C: I remember going to the Speech and Hearing Clinic and they would
let me sit on the other side of this window—you know, *I* could see in but
it was a mirror on the other side. And they were in there working with
Pat. At that time, I was working full time here at home . . . and I had full-
time help. I had a *wonderful* lady who was here with me. Her name was
130 Mary. And she was just like a second mother . . . so, Mary would always
be here. One of the things Mary did for me was that she set the table and
she cooked dinner every night. She left dinner on the stove when she left
at 5:00, and I heated it up and served dinner and washed the dishes. Well,
they asked Pat to put these pictures in an order of what her mother does.
135 And the pictures of cooking and setting the table, she left over here on the
side. And she had the family eating together and the mother washing
dishes. They came back to me and said, "Well obviously this child can't
sequence *anything* because she doesn't know that you set the table and . . ."
And I'm saying to these people, "Wait a minute, wait a minute. You don't
140 understand. I have help that does that every day, so Pat's mother does not
do that." And they just *discounted* me—like this is *irrelevant*! She was
supposed to put setting the table and cooking the meal and eating
the meal and washing the dishes in *that* order. And it doesn't matter that
that's not the sequence we use in our *home*, there's something wrong with
145 *her* because she can't do that. I can remember being so frustrated, trying to
say to them, "No, no, no, no, no! She *knows* what we *do*, it's just that we
don't do what *you* think we do. So, look at what *we* do in our home." I
just remember that frustration.

In questioning the validity of this particular test item for her child, Cam unwittingly challenges the authority granted to professionals while simultaneously exposing their supposedly objective and scientifically developed tools as susceptible to cultural bias (Hanson, 1993; Losen & Orfield, 2002), not only in regard to assumptions about how families *should* operate but also the role of women in the family—a role certainly up for debate in the early 1970s. The professionals, invested in their own authority and belief in the sanctity of standardized tests, immediately dismiss Cam's contextual explanation for her child's behavior as irrelevant to the task of assessing what Pat "was *supposed* to do" (lines 141–142). From the vantage point of the present, Cam mocks the self-assured interpretation offered by the examiners, then recounts how she tried to make "these people" (line 139), a derisive reference to the professionals, understand that it *matters* that "that's not the sequence we use in our *home*" (line 144). In the retelling of this incident, Cam asserts the superiority of her own knowledge, harshly critiquing the authoritative assumptions about her child: "There's something wrong with *her* because she can't do that... 'No, no, no, no, no! She *knows* what we *do*, it's just that we don't do what *you* think we do. So, look at what *we* do in our home'" (lines 144–147). Yet at the time of the incident, Cam remembers feeling dismissed, powerless, and frustrated.

Similarly, Mimi recalls how Pete's kindergarten teacher in the early 1970s (prior to PL 94–142) responds to her attempts to collaborate in light of Pete's significant medical and developmental history.

> M: When we finally had him potty-trained, he was
> 215 five years old, I guess, we entered him in kindergarten. I went in and explained to the teacher about *all* of the problems that we had had—where we had come from . . . and I called her *a lot* in the beginning. I began to realize that she did not want me to call her anymore and that I was *bothering* her. So, I quit calling her. We would go for what little things they had and Pete would always be—he would never be participating—he would be sitting over to the side. Very sweet—very, very, very, very, very *sweetly* sitting over to the side. Made *no* problems for anybody. Just quietly sat there.

From this early experience, Mimi learns three important things about the way schools operate. First, a mother's knowledge about her own child is not necessarily considered desirable, valuable, or relevant by school professionals. Second, school professionals prefer to direct the nature and extent of interaction with parents. Third, a parent who initiates contact with the school is at risk of being viewed as bothersome. Mimi soon conforms to these expectations and disengages from contact with the school. However, she begins to suspect that "knowing her place as a parent" might not be the best approach to getting her son educated. When invited into the classroom by the teacher, she notices that Pete "would never be participating—he would be sitting over to the side. Very sweet—very, very, very, very, very *sweetly* sitting over to the side. Made *no* problems for anybody. Just quietly sat there" (lines 214–225). Her suspicions are confirmed in the following narrative that takes place at the end of Pete's kindergarten year.

M: We finished the year and I went up for a conference, she said to me, "Mrs. Bing, I think that Pete is partially deaf. Have you ever thought that?" And I said, "No, ma'am. Absolutely not." And she says, "Well, you know he does not know any of his colors. He does not know any of his numbers.
230 He cannot carry out two-part or three-part commands. I call to him and he does not answer me. The children speak to him and he does not answer them. I really believe that he has *serious* hearing loss." Well, I was *very* angry with her. I didn't *say* anything to her, but I was very, *very* angry because I knew that she had known this a long time *whatever* it was she
235 was thinking and she had been thinking it a long time. She was *just* now telling me about it and that I had lost ground with it and so forth.

Mimi is outraged that none of these concerns has been communicated during the school year, particularly in light of the teacher's earlier rebuke of her repeated overtures for collaboration, and stunned by the teacher's overall conclusion that Pete must be hearing impaired. Yet, Mimi continues to perform her role as submissive parent, admitting that she "didn't *say* anything to her" (line 233) despite being "very, *very* angry" (line 233). Mimi's understanding of how a parent was expected to behave is corroborated by Della's recollection of her own experiences with schools during the early 1970s: "You weren't as free back then to visit the schools as you are now. And you were looked *down* upon if you came to school. [Teachers] just didn't have time to be bothered with you."

On the basis of the intimate knowledge she has of her child, Mimi does not believe that Pete has any kind of hearing impairment whatsoever. Yet, she recalls how she doubts herself when confronted with a professional's differing opinion.

M: I *knew* that he could hear. And so I brought him home and, *interestingly* enough, the hearing thing really was *bothering* me very badly because I already knew the other things he couldn't do. But, the *hearing* thing really
240 puzzled me. I put him out in the driveway and I went upstairs to my room and called him. I said, "Pete? *Pete*!" And he was down in the lower driveway and I *called* him—just to *prove* to myself that he could *hear* [**laughter**]! And he turned around! "Come in the house!!" He came in the house. And I thought, "What is *wrong* with that woman?!"

In the retelling of this incident, Mimi reveals just how vulnerable she is to professional opinion—to the point that she performs a simple experiment to *prove* to herself what she already believes to be true. Reassured by evidence that Pete does indeed hear, Mimi is confident now to question the authority of the professional: "And I thought, 'What is *wrong* with that woman?!'" (line 244).

Echoing Mimi's feelings of vulnerability, Cam explains how she experienced being silenced by a professional discourse that assumes superiority over a mother's knowledge.

C: It's almost like so many times that you're standing by yourself and everybody else from the doctors to the
370 psychologists to the teachers, *everybody* is sitting in *judgment* in front of you and telling you what's *wrong* with your child. And you're standing here saying, "Wait! Let me tell *you* what *I* know about this kid!" And *nobody* even cares. It was a *hurtful* experience, it was an emotionally devastating experience. There were times when I came home and it would
375 take me two or three days just to pull *myself* back up by my bootstraps and say, "It's okay, Cam. You're *not* an *idiot.* You *do* know what your child can do. She *can* do these things. It *is* worth fighting for." And not just *cave in*—just because the feeling was to just to cave in to all that and say okay maybe we just need to institutionalize her. I just *can't* fight this anymore.

In recalling her feelings about such interactions, Cam paints a poignant image of vulnerability—a scene in which she metaphorically sees herself made to stand alone in front of professionals while they declare her child to be unequivocally defective. In the face of this pronouncement, Cam envisions herself frantically shouting into unresponsive silence, "Wait! Let me tell *you* what *I* know about this kid!" (line 372).

Authoritative imagery courses through Cam's narrative metaphor. She stands while professionals sit, much like a naughty schoolchild sent to stand before an imposing principal. She describes professionals as "sitting in judgment" (line 370) about what is wrong with her child while she, in contrast, is left standing there. As in her previous narratives, Cam perceives her interactions with professionals as conflictual rather than collaborative, noting that she sometimes felt that she might "just cave in to all that" (line 378) because she just couldn't "fight this anymore" (line 379). Nonetheless, Cam continues to resist the "normalizing gaze" of professionals that "establishes over individuals a visibility through which one differentiates them and judges them" (Foucault, 1975, p. 184). Her summation of these interactions is described as "a *hurtful* experience; it was an emotionally devastating experience" (lines 373–374). It is of interest that Cam initially describes the experience as "hurtful," then immediately chooses to reword as "emotionally devastating"—possibly because in hearing herself say the first description aloud, she recognizes that it does not nearly communicate the depth of her pain. She recalls that it often took several days to recover and talk herself back into reclaiming her own knowledge—"It's okay, Cam. You're *not* an *idiot*. You *do* know what your child can do. She *can* do these things" (lines 376–377)—illustrating the power of an authoritative discourse to disrupt a mother's convictions about what she *knows* to be true about her child.

Influence of Race/Culture, Social Class, and Gender

Given that this generational group represents both the dominant demographic in America and the typical profile for families at the time with children labeled LD (White and middle to upper socioeconomic level), it is unsurprising that race as a factor in parent/professional interactions is scarcely mentioned within the narratives of these mothers. Furthermore, all of the mothers reside in the southeastern United States where, in some contexts, public discussion of race remains taboo. Only Cosette addresses the possible influence of race upon her experiences.

> C: I think probably because I'm a White lady, probably is to my advantage.
> I think because I am an *educated* White lady is probably even *more* to
> 145 my advantage. Because we were in a good public school setting—with
> the first two children—one of the better school districts in the town
> and then conferences from then on were in private school settings and I
> think *all* of those situations probably gave us an advantage—personally.

Cosette's wording belies a tentativeness in her acknowledgment of the privilege inherent in being a White American. For example, she uses the word "probably" four times (lines 143, 144, 148) as if leaving open the possibility that race may not, after all, be related to privilege. Similarly, she tempers her first two statements by prefacing them with "I think" (lines 143, 144), implying that she is not certain that what she is about to state can be considered absolute fact. Cosette goes on to suggest that social and economic status (which she implies is solely the consequence of education), contributes more significantly than race to any advantages she may have had in parenting children labeled LD (e.g., being in a good public school, attending the better school districts in town, accessing private school settings).

All of these White, middle- to upper-class mothers share the advantages to which Cosette refers, yet their narratives indicate that access to better schools and resources did not necessarily guarantee a "good education" for children labeled as performing "outside the norm." With the advent of PL 94–142 in the mid-1970s, parents are guaranteed the right to participate in educational decision making. In actuality, however, these mothers report repeatedly being positioned as passive by an authoritative professional discourse. In an effort to resist domination and assert their legal right to participate, mothers strategically draw upon resources afforded to them by their social and economic status. Mimi, for example, responds to the school district's denial of her request for a third-party evaluation by calling upon political connections available to her as the wife of the owner/editor of a local newspaper.

M: We went to an attorney here in town whose father was a state senator. I had already learned that you *had to have* people around you that had connections. The school district did
850 not *care* and they would run over you if you were a little weasel. But if you brought *power* with you, they wouldn't turn over immediately but they would *begin* to listen and I had to *continue* to try to bring power. I couldn't have *just* an attorney who knew about the law. I had to have one that didn't know *anything* about the law who was *politically*
855 *connected*. And *I* had to know the law. I had pretty much gotten to know the law pretty well by then. This fellow agreed that he would be our attorney if I would help him. And I did.

Perceiving school personnel to be more invested in maintaining a dominant position than in serving the educational needs of her child, Mimi deems it necessary not only to "bring power" (line 852) to strengthen her argument, but also to "*continue* to try and bring power" (line 852) to *ensure* her rightful position as a participant, despite her legal right to do so. In her decision to approach an attorney "whose father was a state senator" (line 848), Mimi reveals her understanding of school as a highly politicized space—so much so that she believes a "politically connected" (line 854) attorney to be more advantageous to her than an attorney well versed in PL 94–142.

Cam, likewise, operating out of a high social and economic status, draws upon her access to powerful people to support her negotiations with the school district.

C: I had to get very tough and very threatening, which was not in my personality, to *get* somebody to even listen to me and to try to find an answer. They didn't *hear* me before that and they were not willing to look
230 for an answer. They were going to march down *their* path with *their* test scores, and if their test scores didn't fit the state criteria, then that was *it*. There was no left, no right, no margin anywhere. You had to be on this path or it was nothing. I just simply refused to accept that . . . *I* decided at that point that something *had* to be done. So, I started making phone
240 calls. I called professors, I called people on the State Board of Education, I called state legislators, I bugged the school board, I hired attorneys, I threatened to sue the school board—I just made all the noise I could make.

> I talked to everybody I could talk to. I tried to get suggestions from everybody. I finally talked to a state legislator from Harrison who helped
> 245 me. He went with me to meet with the teachers, the principal, and with the psychologist and all that . . . when I called and set up a meeting and a state legislator was going to attend the meeting with me, people started sitting up a little bit more and saying, "What can we do?"

Cam recalls the scientific/legal/educational discourse surrounding her child as unilateral and absolute. The power of this discourse is revealed within Cam's retelling. Her choice of words illustrates the authoritarian stance of these professionals. For example, she describes them as determined "to march down *their* path with *their* test scores" (lines 230–231), a militaristic image of foot soldiers obediently following orders to move along a strategically defined route. It is not an image that connotes collaboration, as further reflected in Cam's repeated emphasis upon the word "their" (lines 230–231). The institution of special education, with its particular rules, language, and practices, appears to Cam as a monolithic authority with the power to define the parameters of possibility with "no left, no right, no margin anywhere" (line 232). By restating this point two more times within her narrative, Cam suggests the significance of facing authoritative professionals who "didn't hear" (line 844). In response to an aggressive discourse with the power "to form a body of knowledge" (Foucault, 1975, p. 220) that objectifies and fixes her child, Cam reports that she had to "get very tough and threatening, which was not in my personality, to *get* somebody to even listen to me and to try to find an answer" (lines 842–844). Cam's agency in responding to this aggressive discourse is reflected in her use of active rather than passive sentences (e.g., "I" appears 15 times within the narrative text) as well as the use of action verbs (e.g., called, bugged, hired, threatened, talked).

Drawing upon her high social and economic status, Cam gains easy access to people possessing a particular social clout (e.g., university professors, members of the State Board of Education, state legislators, members of the local school board, attorneys). Not only is she able to solicit advice from such people, Cam successfully negotiates with a state legislator to attend an eligibility and placement meeting on her behalf. She is well aware that the presence of this politically powerful person will shift the power distribution to open a space for her to be heard: "When I called and set up a meeting and a state legislator was going to attend the meeting with me, people started sitting up a little bit more and saying, 'What can we do?' " (lines 247–249). Thus, it seems clear to Cam that a mother risks being disregarded by professionals unless she is able to bring power to support the legitimacy of her contributions.

Mimi's narrative, set in the late 1970s, corroborates Cam's perception that the parental rights guaranteed under PL 94–142 are not strong enough, in practice, to offset a power distribution that favors professionals.

M: We took a state representative, Leonard Pfeiffer, who, believe it or not, lived right here in this town and who was the chairman of the Education Committee of the House of Representatives. Why? Because that was how high stakes the game was! I mean, we knew the stakes were *high*. And that *some*body who had some *authority* who could *stop* this 200 million pound
295 *gorilla* had to be there to see how *awful* it was. So, Mr. Pfeiffer, Representative Pfeiffer was also a retired educator from Pleasantville County who knew where all the facilities were in the county and knew everything about the county. Interestingly enough, in this letter that our attorney wrote telling who was coming with us, I thought it was real
300 interesting that he did not say that he was a member of the House of Representatives. Of course, they *knew* that. But, he told who he was by saying he was a retired educator 30 years in the school district and all the things that had to do with his school affiliation, of course. It really didn't matter—should not have mattered—but I *knew* it did.
305 We *all* knew it did that he was chairman of the Education Committee of the House of Representatives. Anyway, we went down there for it. It was very *cold*. You could have cut the air with a knife, it was so cold [**speech slows for emphasis**]. It was just *horrible*.

In an effort to redistribute the power relations in her ongoing negotiations with school district personnel, Mimi relies upon her social and economic status to engage the chairman of the Education Committee of the State House of Representatives on her behalf as "somebody who had some authority who could stop this 200 million pound gorilla" (lines 294–295). In likening the school district to an unstoppable beast, Mimi offers a metaphor to emphasize just how significant she perceives the power differential to be—as significant as King Kong blithely crushing any and all obstacles in his way while the powerless in size and strength frantically run in all directions.

Mimi, like Cam, understands the need to bring power to the table because "that was how high stakes the game was!" (lines 292–293). It is of interest that Mimi calls the process of negotiation a game of high stakes—perhaps an unconscious inference to the status of people who typically engage in high-stakes activity. The game-like metaphor continues in Mimi's description of how her attorney metaphorically plays his cards by holding back the trump everyone knows he holds: "He did not say [Pffeifer] was a member of the House of Representatives. Of course they knew that" (lines 300–302). She recalls the negative response to her persistent efforts at restoring a balance of power, noting that "you could have cut the air with a knife, it was so cold [**speech slows for emphasis**]. It was just horrible" (lines 307–308).

The narratives told by this generation of mothers, situated within the historical time frame of the 1960s through the mid-1980s, reflect larger cultural assumptions about women, in general, and mothers, in particular. Despite the introduction of feminist discourse into the nation's cultural consciousness (and more specifically the workplace), these narratives suggest that school professionals, mostly women, continue to hold traditional attitudes toward mothers. Although the roles of women (and mothers) expanded greatly during this period, school personnel continue to define responsibilities of motherhood in traditional ways (Cutler, 2000; McFall, 1974; Reid & Valle, 2002). For example, all five mothers report that school personnel saw them as bearing sole responsibility for their children's educational well-being. None of the mothers could recall an instance when school personnel contacted her husband about a school issue. Mothers claim that school professionals, operating out of a patriarchal stance, not only dictated *how* they should enact their responsibility but also blamed them (never their husbands) for the failure of their children. As Della explains, "It was *all* on me and I felt I had nowhere to go, nowhere to turn."

The following narrative gives insight into the degree to which mothers of children labeled "deficient" may feel judged in regard to the cultural ideal of The Good Mother—that is, "good" mothers raise successful, above-average children (Warner, 2005). Mimi, faced with school professionals who observe her child's week-long evaluation at a university child development center, describes her feelings as she dressed herself and her child each morning.

M: And I can remember that morning getting up with
Pete at the motel to go. Because you see we had to stay in Wellford
105 the whole time. I can remember getting up and getting ready to go and
getting *him* ready and thinking *every* time I put something on him
[**voice lowers**], "Is this the *right* thing? If he wears *this*, will they think that
I am a prissy mom that is spending too much time *dressing* my child and if
I have my priorities in the *wrong* place?" I thought about what *I* wore and
110 how I was going to present *myself* to them and was it going to be *wrong*?
You know, you're damned if you do and you're damned if you don't. And
I *was* a very particular mom. Pete had lovely clothes . . . So, I can remember
that day being very, very concerned about—well, *every* day that we went
for testing—about what he had on.

If we were to retell this narrative by substituting a father in place of the mother, the narrative surely would fail to make any cultural sense to us whatsoever. This kind of judgment (even the perception of such judgment) is reserved for mothers only. Such is the entanglement between mothers and their children. And as Mimi remarks, "You know, you're damned if you do and you're damned if you don't" (line 111).

The cultural expectation for women to be accommodating rather than assertive is clearly seen in Cosette's description of requesting classroom accommodations for her daughter: "I tried to be very nice and ladylike, but I'm not sure that I always was. So, I was probably a pain in the neck [**light laughter**]!" Despite her right *under the law* to make such a request, Cosette exercises that right in a way that conforms to cultural expectations for women to behave in very nice and ladylike ways—in other words, to be compliant, uncritical, and accommodating to others (Belenky, Clinchy, Goldberger & Tarule, 1986; Code, 1991). Admitting that she may not have managed to always maintain a demure demeanor, Cosette appears to accept, rather than reject, the cultural perception that assertive women are somehow problematic (unlike assertive men), when she states, "So, I was probably a pain in the neck [**light laughter**]!" Perhaps Cosette's light laughter belies some degree of tension beneath her statement.

In relating her experiences interacting with school professionals, Lindsey recalls a striking difference in professional treatment, depending upon whether or not her ex-husband attends the meetings.

> L: By then, I was by myself and I was divorced. But I was getting smarter. I realized that when I went to meetings—even though my husband never said anything—if he was there, I got something going on as if the principals were intimidated by a man. And he honestly didn't say anything. He just came and sat in the room. But when he was not there, they were treating me like an hysterical woman.

Suspecting that professionals perceive her as irrational by virtue of her gender, Lindsey solicits her ex-husband to accompany her to subsequent meetings. The mere presence of her ex-husband shifts the power distribution enough to open a space for Lindsey to speak and be heard. Given that her ex-husband contributes nothing in the meetings other than his presence, Lindsey's narrative points to a cultural standard that grants men (White men in particular) both rationality and power on the basis of their embodiment (Young, 1990).

Much like Lindsey, Mimi's lived experiences as a woman and mother substantiate her understanding that men possess power and the only way for women to secure power is to align themselves with the men who have it. Certainly, Mimi's attempts to exercise her rights under PL 94–142 validate this belief. In response to having had her knowledge repeatedly dismissed by school professionals, Mimi solicits male power (e.g., attorneys, political figures) to speak on her behalf (Shildrick, 1997). Thus, it is unsurprising that Mimi, dissatisfied with what she sees as persisting educational inequities for children with disabilities, asks the men in her family to run for school board in hopes of gaining access to the power needed to impact public education.

M: I came and I asked my daddy to run for school board and I asked my husband to run for school board and I asked my brothers and I asked other people that I trusted would get on the school board and help me
390 take care of handicapped children. Everybody said no. One of them said, "Well, why don't *you*?" And I said, "Oh, I could *never* do that. It's not something that I could ever do."

Having internalized cultural expectations for the role of mothers, Mimi, a self-described "professional mom," immediately dismisses any notion of running for school board herself. Yet after careful reconsideration, Mimi decides to take the risk. Following a successful campaign, she wins a seat on the predominantly male school board. After having raised her status from full-time mother to school board member, Mimi reflects upon the dramatic shift in her relationship with school professionals.

M: Anyway, Pete finished middle school and during that time I was a member of the school board. Things were different. Considerably different. I still had exactly the
715 same zeal for him, but I got treated differently. I mean, it was like a difference in night and day. It was really unbelievable. I did not change. But, it all changed.

Mimi's narrative provides compelling evidence for the low status that school professionals routinely assign mothers. Prior to her election to the school board, Mimi feels compelled to strengthen her low-status position as a mother by bringing "male power" to negotiations with the school district. Once elevated to the status of school board member, Mimi suddenly commands the kind of attention afforded to the men who once spoke for her. As she points out, "*I* did not change. But it all changed" (lines 716–717).

Conclusion

During this historical time frame (1960s to mid-1980s), a discourse of special education develops and emerges within the institution of public schooling. With the passage of PL 94–142, the discursive practices of medicine and law merge with education to create new ways to speak about and act upon children with disabilities. Drawing upon its parent disciplines of medicine and psychology, the institution of special education gives rise to an apparatus through which a medicalized discourse of disability can circulate. The new law guarantees parents of children with disabilities the right to participate in educational decision making; however, mothers of this generation document that professionals, operating within this new and hybrid discourse, claim an authoritative stance that precludes "other ways of knowing" as well as meaningful parent/professional collaboration. On the basis of their experiences within the institution of special education, mothers of this generation report that professionals dismiss, in particular, the kind of contextualized and subjective knowledge parents bring about their children in favor of scientific knowledge gleaned from objective and decontextualized assessments. In response to being passively positioned by school professionals, these mothers (White and middle to upper class) strategically draw upon their cultural collateral (social class, material resources, and male power) to gain rightful access to the process.

Chapter Four

The Implementation Years: Second Generation Mothers

I am defining this second historical time frame as the years between the mid-1980s and mid-1990s in which implementation and compliance become the focus of special education, an era in which both schools and parents test the limits of the law. In this chapter, I include two mothers introduced in chapter three, Cosette and Lindsey, whose later narratives cross into this time frame (see appendix E). I introduce five mothers whose children, born in the 1980s (with the exception of one older sibling who does not figure extensively into her mother's overall narrative), were of school age during this period and labeled learning disabled (LD): Rosemary whose daughter Rebecca was born in 1981; Marie who has a daughter, Tia, born in 1979 and a son, Gregory, born in 1982; Chloe who has two daughters, Jessie born in 1982 and Ashlyn born in 1988; Sue whose son Evan was born in 1984; and Dawn who has a son, Shawn, born in 1985 and a daughter, Brittany, born in 1988 (see appendix B). All of the mothers are White, but represent a range of socioeconomic level. Two mothers identify as middle to upper class, two mothers identify as middle class, and one mother identifies as lower-middle class (see appendix C).

It is noteworthy that only two participants chose to use pseudonyms for themselves and their family members. The remaining three mothers rejected the use of pseudonyms, choosing instead to reveal their identities. These mothers describe their identity disclosure as a political act—an opportunity to name their experiences aloud.

Participant Snapshots

Following the format of chapter three, I begin this chapter with participant snapshots, or brief biographical sketches, to acknowledge each mother as an individual with a particular story to tell as well as to orient the reader to the general positionality of each participant. I also describe the nature of my relationship with each mother.

Rosemary

I first met Rosemary when she brought her youngest child, then six-year-old Rebecca, to be evaluated at the developmental pediatrics clinic where I worked as an educational diagnostician/consultant. At the intake conference, Rosemary, a former teacher and stay-at-home mother to Rebecca and her three older and academically gifted siblings, expressed confusion and concern about Rebecca's persisting difficulty in acquiring basic academic skills. As a member of the clinic's multidisciplinary team, I administered Rebecca's first educational evaluation, thus beginning a consultative relationship that would span more than a decade and ultimately encompass all of Rebecca's elementary, middle, and high school years.

As I walked alongside this family through an educational system seemingly designed for certain students and not others, I acquired intimate knowledge of the struggles that bonded mother and child in a shared resolve to scale new obstacles just as old ones receded. Rosemary refers to the unrelenting pressure as "the merry-go-round [that] just keeps going and going. It doesn't stop. You never get off... I *never* think that we're climbing a hill and I can see the top. You just go round and round." Yet, at no time on this journey does Rosemary waver in her devotion to her daughter's education. As Rebecca gains academic skills, Rosemary simultaneously gains advocacy skills. In response to the many teachers who doubt Rebecca's potential, Rosemary insists upon equal opportunities for her daughter. And time and time again, Rebecca proves the naysayers wrong. Year by year, mother and daughter persist in the agonizing passage through school.

Rosemary's narrative illustrates the costs of such intense and ongoing engagement in a child's schooling, including the eventual unraveling of her marriage and her temporary relocation to another city so that Rebecca can attend a private high school for students labeled LD. Rosemary remains philosophical about the changes in her life, ultimately satisfied with the choices she made and who she has become. Rebecca now attends college. Rosemary admits, however, "I'm *tired.* I'm tired of having to help this child fight her battles. It does take someone *else.* But this has been going on now since she was four years old. And, she's 22. And we're not through with education yet."

Marie

All of my life I have reminded people of other people they know. Thus, when a colleague told me that I had to meet her friend who reminded her exactly of me, I dismissed her enthusiastic insistence about our supposed similarities. At a subsequent social event, this colleague introduced me to her friend Marie, leaving us alone to discover whatever uncanny resemblance that she saw in us. After a mere 30 minutes or so of conversation, we commented in astonishment, "It's like looking into a mirror!" Not only do we physically resemble one another but our career paths have unfolded in remarkably similar ways. We quickly became both personal and professional friends.

Marie begins her narrative within her own childhood. As the sister of three very bright yet chronically underachieving brothers, Marie, a self-described "good Catholic school student," recalls the nuns repeatedly remarking, "Oh, you are so different from your *brothers*!" Puzzled by the discrepancy between her brothers' evident intellectual gifts and their erratic academic performance, Marie eventually pursues a college degree in psychology, in part to better understand the reasons for her brothers' difficulties in school. Initially suspecting sex role differences in the family as a contributing factor, Marie is taken by surprise when her psychology professor suggests that perhaps her brothers have undiagnosed learning disabilities, a condition that she has never heard or read about. From that time (early 1970s) forward, Marie's personal and professional life becomes closely entwined with the emerging diagnostic categories of learning disabilities and attention deficit disorders.

With degrees in psychology and social work, Marie establishes a thriving private practice for families of children with learning and attentional issues. Marie is, nonetheless, taken by surprise when the now familiar "symptoms" emerge within her own family, manifesting to a mild degree in her daughter and to a far more significant degree in her son. Marie's narrative, from her viewpoint as both a professional *and* a parent, traces her struggles to find and maintain educational environments she deems appropriate for her children.

For nearly two decades, Marie devotes herself not only to the education of her children but also to families, like her own, who struggle with multiple issues related to learning disabilities and attention deficits. She credits her children for providing her with wisdom and strength: "I think when I look now at my children in their early 20s versus who I was in my early 20s, they are much stronger. Much stronger. Gutsier. Survivors [**pause**]. I've become a survivor through them."

Chloe

I have often told Chloe that she is in the world like bottled sunshine. It is impossible not to feel good in her presence. When I think of Chloe, I see her dressed in the trademark vibrant colors and patterns that announce her vivacious presence. I first met Chloe, a stay-at-home mother of three, when she brought her teenage son, Rob, for an evaluation at the developmental pediatrics clinic where I worked as an educational diagnostician/consultant. Although he displayed a discrepancy between his verbal and written expression, Rob did not meet the school district's eligibility criteria to receive learning disability services.

Rob's evaluation and follow-up conference, however, mark the beginning of Chloe's understanding of learning disabilities. She soon recognizes "the signs" in her youngest child, then six-year-old Ashlyn, who struggles to acquire basic academic skills, and later in her middle child, Jessie, who fails high school mathematics. Once she enters the realm of special education, Chloe finds interactions with professionals to be highly unpredictable—positive in one instance and devastating in another.

Inspired by the struggles of her children, Chloe becomes PTA president for Ashlyn's middle school in hopes of raising awareness about children with learning difficulties. In her position, Chloe gains intimate knowledge of the ways race, class, and ability intersect with educational opportunity. Distressed by what she sees and hears, Chloe commits to being a full-time volunteer tutor for children with learning difficulties in hopes of somehow equalizing educational opportunities for all.

Chloe's passion and outspokenness about her children's academic difficulties, as well as those of other people's children, is not without consequence. To her established, well-to-do in-laws, Chloe violates the long-standing code of Southern White families that decrees collective silence in regard to family matters—particularly any matters (e.g., disability) perceived to cast aspersions upon the family name. Such tensions become a contributing factor in Chloe's now impending divorce. However, like Rosemary, Chloe is satisfied with her life choices and continues her mission to bring attention to all children who struggle to learn.

Sue

By seven years of age, Evan had managed to acquire a checkered school history. Not only had he been asked to leave preschool because of behavioral problems, but he also had amassed a laundry list of teacher complaints in his brief elementary school career. Sue, an artist and owner of an advertising agency, knew a different Evan. "With me, he was engaging, he was funny, he *loved* to draw, liked to build stuff, but he had his own little drummer. And I just thought, you know, well, that's just the way it is. I'm a creative mother and that's just the way it is." Yet, when school personnel suggest an evaluation, Sue dutifully brings Evan to the developmental pediatrics clinic where I worked at the time as an educational diagnostician/consultant. Results of the multidisciplinary team assessment indicated mild learning disabilities and attention deficits; however, Evan's wildly inventive manner of being in the world (coupled with a wit far surpassing his years) seemed to be the more salient and significant aspects of this engaging little fellow. I had difficulty reconciling the troublesome, academically delayed student described by his teachers with the intensely creative little boy with whom I instantly became enamored.

Evan's innovative nature would remain at odds with traditional academic expectations throughout his schooling. During his early academic years, Sue, a divorced, working mother, struggles to balance the demands of career and single parenting. The ordinary challenges of single parenting intensify with Evan's increasing difficulties in school. Daily homework completion feels to Sue like "another *job*. Another *gut-wrenching* job." Although she remarries after a few years, Sue remains solely responsible for negotiating Evan's education.

Thinking back upon Evan's school years, Sue sees herself and her son together on a lone journey against the world—fighting misconceptions and assumptions held by educators, administrators, and extended family as well as her ex-husband. "You feel like... there's something seriously wrong because you gotta *make* people look at him for what he *is* and what he *can* be, not what's on the paper." After years of exclusion, in one form or another, within traditional classrooms, Evan now attends the Savannah College of Art and Design where he is flourishing. Sue muses upon Evan's success in college: "It's all of sudden letting him be *himself*."

Dawn

Unlike the long-term relationships I have with the other mothers in this generational group, Dawn and I met for the first time on the day of the interview. Dawn was referred as a potential participant by my sister Pam Carter who collaborated with the family in her former position as a resource teacher for two of their four children. Upon learning more about the nature of my study, Dawn eagerly agreed to participate and arranged a date for the interview.

On the day of the interview, I follow detailed directions to Dawn's home in a rural area of a southeastern state. I turn onto a barely noticeable gravel driveway that meanders through the family's property, passing by tool sheds, a large building used by Dawn's husband for his machinery repair business, and the usual big equipment used for the upkeep of extensive acreage. Barking dogs announce my eventual arrival at the house. Dawn greets me with a broad smile. Immediately taking note of similarities between my sister and myself, Dawn comments that she feels like she already knows me. As a "child of the South," I breathe in the warm familiarity of the rural environs and Dawn's gestures of Southern hospitality. We settle comfortably into the interview.

Dawn begins her narrative at the time her second child, then seven-year-old Shawn, qualifies for learning disability resource services. Initially frustrated by words and procedures she does not fully understand, Dawn starts the process of educating herself about learning disabilities, and in doing so, discovers the source of her own struggles in school—"I didn't even know *I* had learning disabilities until my child was diagnosed with it!" Shortly thereafter, Brittany, her third child, is also identified as having learning disabilities. Dawn notes, "That was just the beginning of my education to deal with what *I* have and helping them to understand what *they* have. That's how it started."

As a stay-at-home mother whose children attend charter schools, Dawn volunteers much of her time in the classroom. She attributes her generally positive relationships with teachers to her presence at the school as well as to the more informal atmosphere that a charter school offers. As a result of working closely with teachers, Dawn feels more confident to support the academic needs of her own children.

On the basis of her experiences, Dawn believes that parents need to be educated about learning disabilities so that they can better help their children and perhaps, as in her own case, help themselves. Dawn and her children support one another in moving toward their shared goal of improved reading skills. Dawn explains, "Reading was always a struggle for me. But now that I have started doing college courses and I've been having to read two or three chapters a week, it's like overwhelming. Now I can *read*!"

Second Generation Mothers: A Collective Narrative

In the following section, I present a collective narrative crafted from the mothers' individual stories that take place between the mid-1980s and mid-1990s. The experiences reported by these mothers reveal their perceptions of the power distribution within special education discourse, how they position themselves within and against special education discourse, and the consequences of special education discourse in their lives. The stories are grouped into four broad themes: (1) the language of experts; (2) conflicts in shared decision making and individual education plan (IEP) implementation; (3) devalued knowledge of mothers; and (4) influence of race/culture, social class, and gender.

The Language of Experts

The mandates required by the Education for All Handicapped Children Act (PL 94–142), reauthorized and renamed in 1990 as the Individuals with Disabilities Education Act (IDEA), establish special education as *the* socially agreed upon institutional response toward educating American children with disabilities. In turn, professionals (e.g., administrators, teachers, school counselors, school social workers, school psychologists) adopt the "language of special education"—a *particular* way of talking about and responding to struggling learners—a hybrid language constructed from the merging of discourses (scientific, medical, legal, bureaucratic, and educational) that constitute special education practice. This new professional language, complicated further by the field's prolific use of initials as shorthand for special education terminology (e.g., IEP, FAPE, LRE, IDEA), becomes the taken-for-granted mode of communication for speaking about children with disabilities in public schools.

As a part of the strategy for implementing PL 94–142, school districts routinely offer mandatory workshops for school professionals in a systematic effort to circulate the discourse of special education. In contrast, parents have far less exposure to and/or practice with the specifics of the new law and its accompanying new language. Thus, as school personnel begin to speak and act in ways consistent with special education discourse, it is unsurprising that some parents respond with confusion, passivity, and/or suspicion. Nor is it surprising that, given the entrenched pattern of public school domination over parents, a knowledge disparity persists between professional and parent.

In a narrative set in the mid-1980s, Marie recounts her initial introduction to the discourse of special education. Tia, her oldest child, enters first grade as a curious and exuberant six-year-old. A few weeks into the school year, Marie receives a call from the teacher to solicit her help in getting Tia "to stop asking so many questions" because she "understands the material and she's pushing us ahead and quite frankly, we are not ready for that and I have to teach to the group." Marie, still reeling from this interaction with Tia's teacher, is stunned to receive yet another call a few weeks later with news that Tia is now having difficulty in school.

M: The teacher said,
"She is having *some* trouble learning." And you know, we made jokes!
"Well, what? Did she stop asking questions [**laughter**]? What's the
120 problem here?" But this teacher said that Tia didn't *want* to read and
didn't *want* to be forming her letters. And she knew that the child wasn't
an oppositional child, but she was really looking to get out of it and not
enthused and really wasn't keeping up with what everybody else was
doing. And we should have her tested. So, I said, "Okay." Not really
125 realizing what I was getting into or the road we were beginning. At that
time, I had had psych testing as a student. But it was always looked at
like something you did not for how a student *learned* but how it fit into
their mental health picture. So I was a little put off. "Do you think she's
mentally ill [**laughter**]? What's going on here?" So we had her tested.

In Marie's recollection of this conversation, the teacher clearly stakes out an authoritative position. The purpose of her call is to *inform* Marie about what she has observed about Tia in the classroom, namely, that she "is having *some* trouble learning" (line 118), "didn't *want* to read and didn't *want* to be forming her letters" (lines 120–121), "was really looking to get out of it and not enthused" (lines 122–123), and "really wasn't keeping up with what everybody else was doing" (lines 123–124). At no point in the conversation does the teacher ask for Marie's opinion or knowledge about her own child or engage her in any way. Nor does the teacher offer any explanation for her dramatic change in perception regarding Tia's classroom performance. She concludes with the directive that Marie "should have [Tia] tested" (line 124), her sense of authority reflected in the use of the modal auxiliary "should." It is noteworthy that the teacher does not appear to consider contextual factors that might be influencing Tia's ability to learn (including the quality of her own teaching); rather, she immediately engages in the discourse of special education with its focus upon the "individual as the unit-of-analysis" (Reid & Valle, 2004). Because the institution of special education *exists*, the teacher freely accesses the assessment procedure that sorts struggling learners, thereby negating the immediate need to exhaust all possibilities within the educational context.

In the retelling of this event, Marie inserts what she was really thinking but does not say to the teacher. For example, in response to the teacher's assertion that Tia is experiencing difficulty learning, Marie mocks the teacher's authority: "And you know, [my husband and I] made jokes! 'Well, what? Did she stop asking questions?'" (line 119). Yet at the time, Marie does not challenge the teacher's position and agrees to have Tia tested. From the vantage point of the present, Marie acknowledges that she complied without "really realizing what I was getting into or the road we were beginning" (lines 124–125), reflecting an uneven distribution of knowledge between mother and teacher.

It is noteworthy that this narrative takes place a full decade after the passage of PL 94–142, yet competing discourses persist about children with learning and/or behavioral challenges. Drawing upon her professional training in psychology, Marie understands psycho-educational testing as one of several tools used to construct a profile of an individual's "mental health picture" (line 128). Thus, she feels "a little put off" (line 128) at what appears to be an insinuation that her daughter could be "mentally ill" (line 129). In contrast, the teacher, operating out of the discourse of special education, understands psycho-educational testing as a tool used to explicate a child's learning profile by confirming or ruling out the presence of disability. Marie's confusion is reflected in her plaintive question, "What is going on here?" (line 129).

As Fairclough (1995) reminds us, competing discourses result in the "creation and constant recreation of the relations, subjects, and objects which populate the social world" (p. 73). With the passage of PL 94–142, the category of learning disability emerges as the dominant discourse to explain bright children who struggle to learn, thereby creating new ways of speaking about and responding to children who previously would have been dismissed as underachieving, unmotivated, and/or lacking in ability. Yet, acceptance and integration of a new discourse, even one mandated by public law, is neither a simple nor swift process. Ten years after the passage of PL 94–142, Chloe recounts how she first responded to the discourse of special education.

10 C: I had never been associated with anything to do with LDs before. I didn't know *anything* about it. Nothing. And I figured—like some people that—oh, gosh, this is a bad label, this is a mental thing. And of course, when you call up your insurance company that's what they tell you. This is a mental disability, you know, and we either cover mental
15 disabilities or we don't. So right off the bat, you're put behind the eight ball with, you know, you've got a kid with *mental* problems.

Although public school services for students labeled LD had been in place for ten years, Chloe attributes her lack of knowledge to never having been in a position to "be associated with anything to do with LDs" (line 10), suggesting that the discourse of learning disabilities circulating in public schools had not yet become accessible and/or familiar to persons outside that arena. A full decade after public school personnel adopted this particular way of thinking about and responding to capable children who struggle to learn, Chloe's lack of exposure to this discourse leads her to think that "this is a bad label, this is a mental thing" (line 12). Having chosen to have her child evaluated at a private center rather than through the public schools at no cost, Chloe recalls that her insurance company's classification of a learning disability as a mental health issue (a practice that persists today) only reinforced her anxieties. Chloe's concluding statement reflects her discomfort with a competing medical discourse that pathologizes learning problems as a mental health issue, a conceptualization not shared by the educational community: "So right off the bat, you're put behind the eight ball with, you know, you've got a kid with *mental* problems" (lines 15–16).

Additional evidence of this persisting lack of intersection between medical and educational discourses appears in Marie's narrative about her then three-year-old son, Gregory. In light of her increasing professional knowledge about learning and attentional problems in conjunction with her older daughter's eventual diagnoses of mild dyslexia and attention deficit hyperactivity disorder (ADHD), Marie suspects that ADHD might be a contributing factor in Gregory's difficult behaviors.

M: Actually, I took him to several different neurologists. The first one said, "Well, yeah, but I'm not going to give you any medication now. See how long you can stand it." Meanwhile
260 I had been doing my research and reading and had come to the conclusion that there might just be some severe cases where medication was warranted. My child was so *unhappy*. He was having rages and tantrums *beyond* the norm. And I was a behaviorist **[laughter]**! I *knew* what to do! None of this stuff worked! So then I went to another neurologist and she—
265 I'm trying to remember why I didn't stick with her—oh, she told me, "Yes, he is. He's *so* severe that he's going to have to be *institutionalized.* I understand you want to keep him, but we need to talk about this. Maybe he is autistic and maybe he is **[pause]** schizophrenic. He's very severe." I *knew* from my readings that he was severe. But I also *knew* that he did not need to be institutionalized.

In describing her experiences with two different neurologists, Marie relates only the words of the professionals (lines 258–259; 266–269). It is noteworthy that she does not represent herself as speaking or responding in either scenario. Rather, the professional pronouncements stand alone as if she were not there at all.

In the first scenario, the neurologist agrees with her suspicion of ADHD, yet offers no intervention beyond having Marie see how long she can "stand it" (line 259). Although she does not confront the neurologist, Marie's retelling reveals seeds of resistance to the authoritative discourse that positions her as passive. For example, Marie acknowledges that she does not rely solely upon professionals, but also does her own research as a professional in her own right. A growing sense of agency is seen in her use of "I" within sentence construction: "I had been doing my research" (line 260), "and I was a behaviorist!" (line 263), "I knew what to do!" (line 263). On the basis of the results of her own research, she disagrees (albeit not openly) with the neurologist's opinion about the use of medication for young children. Moreover, Marie considers her own observations of her child as valid evidence to support her position: "My child was so unhappy. He was having rages and tantrums beyond the norm" (lines 262–263).

In the second scenario, the neurologist also agrees that Gregory displays ADHD symptomatology. In stark contrast, however, this neurologist pronounces Gregory as "so severe that he's going to have to be institutionalized" (line 266) and that "maybe he is autistic and maybe he is **[pause]** schizophrenic" (lines 267–268). On the basis of a single interview and no other contextual observations or team evaluation, a child's future is condemned to an institution. In the space of one sentence, separating a mother from her three-year-old child is dismissed as merely something that "we need to talk about" (line 267). The expert offers "compassion" about the news she just delivered by stating, "I understand you want to keep him" (line 267)—as if consoling a child about the impending loss of a pet—yet Marie has not spoken or been asked to speak. Thus, the neurologist, imbued with the social power granted to her profession, illustrates what Foucault (1973) refers to as "the sovereignty of the gaze...the eye that knows and decides, the eye that governs" (p. 89), a power so confident that it requires no justification. It is particularly noteworthy that the narrative takes place between 1986 and 1987, far beyond the days of routine institutionalization of disabled children and more than a decade after the passage of PL 94–142. The neurologist's lack of knowledge regarding the continuum of public school services available for children with disabilities under the age of five—as an alternative to institutional placement—points to a persisting lack of intersectionality between the discursive practices of medicine and the discursive practices of special education.

Marie does not present her response at the time to the professional's directive that her child "is going to have to be institutionalized" (line 266), only that she recalls dismissing the professional's opinion in favor of her own knowledge: "I knew from my readings that he was severe. But I also knew that he did not need to be institutionalized" (lines 269–270). Given the significance of this exchange, it is interesting that Marie begins the narrative by stating, "So then I went to another neurologist and she—I'm trying to remember why I didn't stick with her—oh, she told me...he's going to have to be institutionalized" (lines 264–266). In the same way that Mimi must work to resurrect the pain of being told that her son was mentally retarded, it appears that Marie buried this distressing interaction into the recesses of memory. I prompted Marie to try and remember her feelings at the moment of this exchange with the neurologist.

M: I remember being really *angry* about her presentation of such a thought. That if you really believed that you wouldn't take more time and explanation and lead up to it. But it was very—I remember it being very
290 *cold*, very clear-cut, blatant. "That's probably where you are going to have to go here." I remember feeling really angry and going *numb* and just picking my child up and saying, "Thank you very much, but I don't think I'm in the right place" and walking out. And then wondering, "What am I going to do now?" My way of coping is to do more *research* [**light**
295 **laughter**] and read and read and read and more research—which I was doing.

Marie recalls feeling "really angry" (lines 287 and 291) in response to the neurologist's "very cold, very clear-cut, blatant" language (lines 289–290)—a scene reminiscent of Cosette's narrative in which she is told that her daughter is brain damaged (chapter three). Like Cosette, Marie questions how a professional could deliver such devastating information without any seeming comprehension of its emotional impact. Cosette and Marie both report becoming immobilized in response to the professional's language. Similar to Cosette's feeling of being shot, Marie remembers "going numb" (line 291). Despite her intense anger, Marie does not confront the neurologist directly. Instead, she responds with the conventional respect reserved for medical professionals ("thank you very much, but I don't think I'm in the right place" [line 243]), before walking out disoriented and wondering, "What I am going to do now?" (line 294).

Marie eventually establishes what will become a long-term collaborative relationship with a trusted neurodevelopmental pediatrician. She recalls the language used by this expert on the first day they met.

M: One of things he had first said to my husband and I at the end of our first session was "So, do I get it?" And we said, "Yeah." And he said, "You really believe I get it?" And I said, "Yeah." He said, "Then let me
340 tell you something. Your son is not a *bad* child [**voice breaks**]. He's not a bad person. He's a wonderful, bright, *good* child that has this thing that's called neurochemical imbalance [**voice breaks**]. And he can't help it. And you're not bad parents. And it's not your fault [**voice breaks**]. And it will get better from here."

In retelling the narrative nearly two decades later, Marie's voice breaks with emotion, reflecting the profound significance of this life event. In striking contrast to the neurologist's authoritative posturing, this expert engages in a reciprocal relationship with Marie and her husband—listening, validating, and offering his partnership. Such wide discrepancy in professional opinion—a recommendation of institutionalization on one hand and the offering of a parent/professional partnership on the other—suggests persisting variation among professionals in the way that they perceive, interpret, and act upon children deemed "outside the norm" in regard to learning and behavior. It is noteworthy that Marie's narrative bears striking resemblance to the narratives of Della, Cam, Mimi, and Cosette (chapter three) in which they recall the various labels that professionals used to describe their children at one time or another (e.g., "mentally retarded," "very disturbed," "brain damaged," "needs to be institutionalized").

In a narrative that takes place in the late 1980s, Rosemary recounts her first feedback conference regarding then six-year-old Rebecca. Having eschewed public school testing in favor of a private evaluation, Rosemary recalls the language used by a developmental pediatrician to describe her daughter.

R: I remember him saying that her intellect was okay, but it was—I've
55 never forgotten that—"It's just the way she's *wired* [**light laughter**]!" I *love* that term! And I just imagine a whole bunch of tangled wires. But

that's an *easy* way to understand. I *love* that and it's not like saying she's [**pause**] *retarded* or something because it's *not*. And it's the *most* confusing thing I've been through in my life.

Rosemary responds positively to the doctor's explanation of Rebecca's learning difficulties ("it's just the way she's *wired*" [line 55]), noting that such a description was "an *easy* way to understand" (line 57). Moreover, Rosemary expresses preference for a descriptive explanation over a diagnostic label, foreshadowing her eventual belief that special education categorization limits children's educational possibilities. Rosemary's uneasiness about the practice of labeling is evident in her choice to seek private testing (over which she can maintain some control) as well as in her assertion that "it's not like saying she's [**pause**] *retarded* or something because it's *not*" (lines 57–58). On the brink of Rebecca's school career, Rosemary already recognizes the power of language to determine a child's educational future.

During the early 1990s, Sue, mother of then seven-year-old Evan, has her first exposure to the discourse of special education. In light of Evan's fondness for his second grade teacher (whom she also remembers as "wonderful" and "structured but in a very pleasant, informal kind of way"), Sue recalls her surprise at being summoned for a meeting regarding Evan.

S: Then, I got a phone call from her to come in
and sit down and talk to her and have a conference because she was
having some problems with Evan. Apparently, he just could not pay
130 attention. He wouldn't stay in his seat. He wouldn't answer when he was
addressed. In the middle of any kind of test or anything, he would just put
the paper down and just wouldn't do it. He refused to write things, copy
things off the board, just wouldn't do it. He would get upset if she even
pushed him a little bit to get him to comply with whatever the class was
135 doing at the time. "But," she said, "he's got such a *sweet* personality, I
think you need to have him tested because I think he might have ADD."

In the retelling of this event, Sue reveals her sense that a power imbalance exists in favor of the professional. For example, the teacher's request for Sue to "come in and sit down and talk to her" (lines 127–128) regarding "some problems with Evan" (line 129) implies a one-sided exchange for the purpose of imparting information. And indeed, the teacher presents Sue with Evan's "symptomatology" (e.g., inattention, noncompliance, out-of-seat behavior, resistance to teacher direction) that she characterizes as falling outside the range of "normal" second grade behavior and indicative of an underlying disability. It is worth noting that this narrative takes place more than a decade after the passage of PL 94–142. Thus, special education practice, with its emphasis upon "the individual as the unit-of-analysis" (Reid & Valle, 2004), has influenced general education for a number of years. Evan's general education teacher draws upon the discursive practices of special education as her primary means for interpreting Evan's behaviors, an illustrative example of how "the power of the Norm appears through the disciplines" (Foucault, 1975, p. 184).

Sue's narrative demonstrates the process by which Evan becomes constructed as the sum of his deficits. Under the teacher's surveillance and documentation, Evan's difficulties take on salience as supporting evidence for a suspected disability. By the time she calls Sue for a meeting, the teacher seems invested in this idea and confident in her recommendation for testing. At no point does the teacher ask Sue to contribute her own knowledge about Evan nor does she appear to consider any other contributing factors (e.g., context of the classroom). Thus, Evan can be described only in terms of deficit language. As noted earlier in this chapter, I also evaluated Evan in my former position as a member of a multidisciplinary team. Thus, I can attest to the striking contrast between the teacher's understanding of Evan as "the sum of his deficits" and our understanding of Evan as a highly creative child manifesting mild learning and attentional weaknesses within the context of a traditional classroom setting—a striking example of the social construction of disability (Corker & Shakespeare, 2002; Linton, 1998; Ware, 2004).

It is worth noting that none of the mothers presented thus far chose to have their children evaluated through the public schools. Despite the law's guarantee of parental rights, including participation in educational decision making, these White middle- to upper-class mothers harbor suspicions about the intentions of school professionals. In an effort to maintain greater control over the use of diagnostic information, they opt for costly private evaluations in lieu of the free school evaluations. Thus, it appears that these mothers intuit Foucault's (1975) notion that "power and knowledge directly imply one another, that there is no power relation without the correlative constitution of a field of knowledge" (p. 27). More than a decade after the implementation of PL 94–142, these mothers believe that a school-generated evaluation may diminish their negotiating power, reflecting their perception of a power imbalance in favor of school professionals who generate, interpret, and control knowledge about their children.

In contrast to the preceding middle-class and middle- to upper-class mothers who obtain private evaluations, Dawn, who identifies as lower-middle class, has school personnel evaluate her children at no cost. Recalling the first IEP meeting she attended in the late 1980s, Dawn describes her introduction to the language of special education.

D: For me, and being the first time and not even knowing what an IEP meeting was in the beginning, I had not a clue

420 what all that stuff was—[**pause**] I was a little apprehensive, I'm sure. I can't quite remember that day. But, I probably walked out of there dazed and confused and overwhelmed...At that particular time, I didn't understand. I didn't really understand a lot of the terminology they were using...To this day, I don't know what they are because I never really
425 quite comprehended them which is some of the problem being a parent.

Dawn's recollection illustrates not only the inaccessibility of professional language typically used in IEP meetings, but also the professional assumption that parents know what to expect when invited to an IEP meeting (lines 413–415). As a result of her participation in the meeting, Dawn recalls that she "walked out of there dazed and confused and overwhelmed" (lines 416–417). Dawn attributes her confusion to not being able to understand "a lot of the terminology" (line 423) used by professionals, an example of how "Truth" becomes "centered on the form of scientific discourse and the institutions that produce it" (Foucault, 1980, p. 131). From the vantage point of the present, Dawn admits, "To this day, I don't know what [the words] are because I never really quite comprehended them which is some of the problem being a parent" (lines 424–425). Dawn's admission is particularly poignant, considering that she now has engaged with special education discourse for more than a decade.

Although she may not understand all of the expert terminology, Dawn recognizes that the way professionals talk about children determines particular educational outcomes.

410 D: Alicia was a very bright child and all these teachers were encouraging her. She was *encouraged* to get into the AG program. Where Shawn [**slight pause**] it was different terminology. You know it wasn't a happy thing like *encouraging* your child. There was all these extra programs and everything like that. But going into LD [learning disabilities]
415 wasn't the same kind of tone.

Given her positionality as the mother of both a daughter labeled "gifted" and a son labeled "learning disabled," Dawn is privy to the ways that professionals talk about and respond to giftedness as well as disability. She compares the celebratory and encouraging manner of professionals regarding her daughter's eligibility for the academically gifted (AG) program with the decidedly less enthusiastic tone about her son's eligibility for learning disability services. Given that school professionals appear to place greater value upon her daughter's giftedness, Dawn suspects that Alicia's education will be qualitatively different from Shawn's "special" education.

Conflicts in Shared Decision Making and IEP Implementation

IDEA guarantees parents the right to collaborate with professionals in making educational decisions regarding their children with disabilities. Specifically, the law ensures that parents have the right to be informed, the right to be knowledgeable about the actions to be taken, the right to participate, the right to challenge, and the right to appeal (IDEA, 1990, 1997, 2004). As highlighted in chapter three the introduction of these legally sanctioned parental rights does not take place without some resistance from public school professionals. Sharing experiences that occurred between one and two decades *after* the implementation of special education law, mothers in this generational group (mid-1980s through mid-1990s) attest to persisting tensions between parents and professionals in regard to collaborative decision making.

Foucault (1980) contends that we should be "concerned with power at its extremities, in its ultimate destinations, with those points where it becomes capillary, that is, in its more regional and local forms and institutions" (p. 96). Marie, in a series of six narratives, illustrates Foucault's point in her detailed rendering of the way power operates within local special education committee meetings. The first narrative, set in the late 1980s, revolves around a special education eligibility and placement meeting in which a hostile exchange takes place between Marie and the professionals.

M: They said, "He is not
eligible for a regular mainstream kindergarten class. We want to put
him in our special needs class." Meanwhile, I'm in the field. I'm talking
535 to the doctor, I'm going to conferences. I have my own clients. I'm
reading. I'm *learning* very rapidly here. So, my husband says, "Okay, that
makes sense. We get it. Goodbye." And I say, "No, no, no, no—excuse
me! What is the make up of the kids in this class? What are the
diagnoses?" I explained to my husband that this could very well be a
540 dumping ground class. There are all different kinds of problems children
have. Is this the only class they have for children that don't fit in the
mainstream? And it pretty much was. They were *really* put off that I was
asking these questions... "Who *are* you to ask these questions?!"
Somebody actually said that to me. My response was, "Well, number one.
I am *the mother* of the child that we are looking to put in, but if that
doesn't satisfy you, I'm a professional in the field and this is what I
550 do... You know, who are *you* to ask me, *the mother,* why I should be
questioning where my son will be and the environment he will be in?"

In striking contrast to her earlier narratives that feature only the authoritative statements made by professionals, Marie represents herself in this account as a character with agency. Marie begins the narrative by stating what an anonymous "they" said (line 532), implying that she and her husband, John, are speaking to more than one professional and that their relationship with them is an impersonal one. The professionals immediately position themselves as "the experts," informing Marie and John of the placement *they* consider to be in the best interest of their child (lines 533–534). Thus, Marie and John are offered a single option to accept or reject. By placing "an offer on the table" to which Marie and John must respond, the professionals establish a "we versus them" tenor. It is worth considering the degree to which this scenario replicates the negotiation process enacted by lawyers—evidence of the influence of legal discourse upon special education practice.

Marie describes the affront she feels at being positioned as having nothing to contribute (lines 534–536), her growing sense of agency reflected in six consecutive "I" sentences. In contrast to John's passive acceptance, Marie asserts her right to engage in the process and asks, "What is the make up of the kids in this class? What are the diagnoses?" (lines 538–539). Marie suspects that the class might be a "dumping ground" (line 540), echoing concerns raised by critics during the 1980s (Hartwick & Meihls, 1986; Poplin, 1988). She remembers the professionals being "*really* put off" (line 542) by her audacity to question them. It is of interest that Marie's narrative bears striking resemblance to Cam's depiction in chapter three of how professionals react when she attempts to participate ("*who* does this woman *think* she is?" [lines 803–804]). Marie responds to being positioned in a passive role by asserting her status as both a mother and a professional in her own right. In the retelling of her narrative, Marie seems incredulous still and wonders aloud, "You know, who are *you* to ask me, *the mother*, why I should be questioning where my son will be and the environment he will be in?" (lines 549–551). By challenging the professionals, Marie unwittingly disrupts the "system of ordered procedures for the production, regulation, distribution, circulation, and operation of statements" (Foucault, 1980, p. 133)— that is, "Truth" as constructed by scientific discourse and upheld by the professionals who ascribe to it. Thus, it appears that the professionals interpret Marie's questions not only as an affront to their professional integrity but also to the discipline and institution to which they pledge their allegiance.

Marie presses on, however, with a request to observe the classroom to make an informed decision about the proposed placement. Despite her legal right as a parent "to be knowledgeable about the actions to be taken" (IDEA, 1990, 1997, 2004), Marie is met with resistance, evidenced in the following exchange between the professionals and herself.

M: I said, "Number one. I need to observe this class." "Well, we don't know. *Now* we have to get consents from all the parents [**in an exasperated tone**]!" "You mean you put children in this class

570 without having parents *see* it first?" "Nobody has ever asked before." I was like, "Well, okay—*I'm* asking. So get your consents and let me know when I can come in."

This immediate resistance on the part of the professionals exposes their reluctance to share decision making with a parent. Countering that a classroom observation will require "consents from all the parents" (line 568), the professionals clearly mean to convey to Marie that her request is both inappropriate and burdensome. Furthermore, Marie's "unreasonableness" is implied within the assertion that no *other* parent "has ever asked before" (line 570). Undaunted by characterization as a difficult parent, Marie insists upon the observation.

M: So my husband and I went in and there before my eyes, I saw children with cerebral palsy, children who were deaf—I can't believe how I'm blanking—two children with Down
575 Syndrome, and several children who appeared perfectly *normal* to me in *every* way, shape, and form *except* that they spoke another language. I thought, "Okay, so maybe I don't understand. I can't *hear* their disability because I don't understand their language." Then, we left the class. There was not *one* child that appeared *anything* like my son…I was told, "Well,
605 this is the *only* class we have for him." And I said to my husband, "He's *not* going there. We are *not* doing that. I don't know what we will do, but not *that*!"

Despite her uneasiness about the recommendation for placement in "a special needs class," Marie is taken aback upon viewing the class composition. Registering disbelief at the spectrum of disabilities, Marie perceives no evident commonality among Gregory and these students beyond "not fitting" into the general classroom (lines 572–573). In response to Marie's challenge to this placement as "the least restrictive environment" for her son, school officials counter that it is the "*only* class we have for him" (line 605), thereby failing to offer the continuum of services required by law.

Like Cam and Mimi a decade before (chapter three), Marie recognizes the status differential that exists between school professionals and parents. Given her dual status as mother and professional, it is of particular interest that Marie's status as a mother appears to *negate* her professional credibility. In an effort to rebalance the power distribution, Marie solicits other professionals to lend their legitimacy to her position.

M: I came into meetings, even though I was a professional, I came in with a parent assister from LDA. I would bring my educational tester, Miriam, and my doctor. They would come if I felt that I had a big battle on my hands. So I had *my* team as well as they

685 had their team. So I wasn't sitting there *me* against six of them. Because, you know, you have your special ed administrator, your principal or vice principal—whoever is coming in, your teacher, the school psychologist, the educational tester, if there's any therapy—the sensory integration therapist, if there was any counseling involved—the school counselor.
690 It was for intimidating for one or even two parents to go in and sit across from this *team.* I went to one or two meetings and then said, "This isn't right. It doesn't work." And that's when I found out about parent assisters. It just kept growing because I was working with Miriam and Jonathan, the doctor, and we were becoming a team professionally. They just always
695 wanted to be there, too. They were so interested. Then, I was the first one to say, "By the way, I hope you are all aware that it is my right to *tape* this meeting." And I always taped all of our meetings. So, that made them, yes, get their defenses up and be very much on guard. But also it made them know *their* law because *I* knew it. I came in and if I didn't know it, I had
700 the person with me who knew it. Oh yeah, I had a special ed lawyer [**laughter**]. I started developing the relationship. I didn't know if I'd ever need to bring him in, but I developed a professional relationship with him so I could drop his name on the table, at times.

Marie understands "the right to participate" (IDEA, 1990, 1997, 2004) as an exercise in defense rather than collaboration. In retelling her experiences, Marie draws upon language that suggests conflict rather than cooperation. For example, she recalls how intimidating it felt to "sit across from this *team*" (lines 690–691), a spatial arrangement conveying distance and authority. Marie names each professional (e.g., special ed administrator, principal or vice principal, teacher, school psychologist, educational tester, sensory integration therapist, school counselor) as if to underscore the degree of intimidation inherent in this arrangement. Indeed, it appears that the professionals are poised to defend their particular "regime of truth"—the kind of "Truth" Foucault (1980) sees as "linked in a circular relation with systems of power which produce and sustain it" (p. 133).

Cognizant of the power relations that support and reinforce the professional "regime of truth" (Foucault, 1980), Marie recruits a professional cadre of her own. She explains, "So I had *my* team as well as they had their team. So I wasn't sitting there *me* against six of them" (lines 684–685). Moreover, Marie strategizes a plan to enhance her negotiating powers. For example, she is well aware that her position can be strengthened by exercising her right to tape record the meeting. She acknowledges that her decision to tape record "made them, yes, get their defenses up and be very much on guard" (lines 697–698), reflecting a strategic move on her part to position the professionals on the defensive. She further strengthens her negotiating power by developing a relationship with a special education lawyer to be able to "drop his name on the table, at times" (line 703).

Marie realizes that a working knowledge of special education law is a must if she is to advocate successfully for her son. As a result of her independent study of the law, Marie learns that a child with disabilities who attends private school is eligible to receive special education services, excepting transportation, from the public schools. Having rejected the only "special needs" class offered by the school district, Marie investigates the private sector and finds a classroom suitable for Gregory. The following narrative represents Marie's recollection of the meeting in which she seeks approval for special education services to supplement Gregory's private education.

M: In the beginning, I guess
the first fight was—I'm making the choice to put him in this private
705 school, but I want services from you. And they didn't want to do that
because you're paying to go there, so let *them* do it. You know? But it
didn't work that way. I knew my law. I remember—they were *really*
going to say no and not just approve it. I then turned to—there were
two people—it wasn't the superintendent, but the head of special ed
710 for the district, whatever his title was and then the woman that I had
dealt with all along who I knew was *tough* but I also read a softness
in her. I felt her *protecting* her softness which I didn't get from the
teachers or anybody else. But I *felt* her protecting her softness, so
I felt like that was her vulnerable spot. So when I had gone through
715 everything I had go through to present the case to everybody about
why this child needed services provided by the school system, I simply
turned to her and said, "After everything you've heard about my son,
can you do me a favor? Close your eyes for a minute and imagine he's
your child. Step out of your role as—and I said whatever her title was—
720 and be his mother. And then open your eyes and tell me that he can't
have these services."

Marie's narrative is useful for highlighting the multiple and overlapping discourses (e.g., legal, institutional, scientific) that drive special education practice. In making her appeal to the school district for special education services, Marie not only draws upon IDEA's legal grounding to justify the request, but she also must engage in the "discourse of law" to *be able* to ask for services. Not unlike a lawyer, Marie comes to the meeting prepared to "present the case to everybody about why this child needed services provided by the school system" (lines 715–716). Among those present at the meeting are "the head of special ed for the district" (lines 709–710) and a woman whose role is to "make the final decision" (line 724), both entrusted to act on behalf of the institution of special education. Thus, Marie must negotiate the "institutional discourse" used by special education administrators to implement the law. Although the specific content of Marie's presentation is unclear, it most likely reflects the usual kind of scientific justification required for special education placement (e.g., child observations, cognitive assessment, achievement tests, social maturity measures). Thus, Marie draws upon a third discourse, the "discourse of science," to argue her case for Gregory's need of special services.

The professionals depicted within this scenario maintain the objective demeanor expected from those who work within law, science, and various other institutional frameworks. Marie describes the woman administrator as "*tough*" (line 711), a trait consistent with a professional persona; yet, she is able to "read a softness in her" (lines 711–712). By appealing to the administrator as a woman and a mother, Marie blatantly transgresses the socially agreed upon boundary between professionals and parents. In the narrative that follows, Marie recalls the response to her transgressive act.

M: And the *anger* that came from *every*body else around the table on their team—*except* her because her eyes welled up with tears—that I would be so manipulative because they knew she was in charge. She was going to make the final decision. They *just*—you could
725 feel it—nobody said a thing, but you could hear chairs move, and the tension, and shoulders square, and the eyes narrow and people look at me. I just felt it *all* coming and letting my energy deflect it, I just never took my eyes off her. And her eyes just fell down. And I said, "So, he'll have the program [**voice lowers**]?" And she said, "Yes. You've made a good
730 case."

Marie's passionate plea disrupts the sanctioned discourse and "system of ordered procedures" (Foucault, 1980, p. 133) for special education meetings, evoking anger from the remaining professionals. By publicly challenging this intellectualizing of one small boy's education, Marie indicts a system that privileges law, science, and institutional frameworks above what *feels* like the right thing to do. The administrator's emotional response stands out against the context of a discipline that relies upon, like all other disciplines, "a set of propositions which are scientifically acceptable, and hence capable of being verified or falsified by scientific procedures" (Foucault, 1980, 112). Discomfort permeates the room as the administrator contemplates her forced choice between rationality and compassion.

In the eyes of her colleagues, the administrator's compassionate response reflects weakness—after all, she allows herself to be manipulated by a mother. The professionals around the table appear unable to read this exchange in any other way and respond with anger toward this errant mother. In the same way that Cam experiences collaboration as an exercise in deflecting the language of professionals (chapter three), Marie deflects the anger directed at her by the other professionals. The administrator concedes to

Marie by stating, "You've made a good case" (lines 729–730)—an example of the influence of legal discourse upon special education practice. Later in the interview, Marie muses upon her negotiations with school personnel: "But it was [**pause**] always done as if they were doing me a *favor.* Not as if this is your *right.*"

The preceding narrative well illustrates the circulation of legal discourse (among other discourses) within eligibility and placement meetings. In the following narrative, Rosemary reveals the influence of legal discourse upon classroom practice.

R: In third grade
was when we *really* knew things were *not* going to work. And the
teacher kept saying, "Rebecca fails the test. But when I bring her to the
35 back and give it to her orally, she tells me everything. So, I know
Rebecca *knows* this material." But, the fourth grade teacher would *not* do
anything oral unless it was written up because she said she could not
do that unless it was a 504 Plan.

In the first scenario, the third grade teacher concludes that oral testing is the most reliable measure of Rebecca's knowledge, thereby exercising her professional discretion to decide what counts as knowledge in her classroom. In contrast, it appears that the fourth grade teacher sees her professional discretion restricted by a legal discourse that requires justification for oral testing as a classroom accommodation under Section 504. However, it is also possible that the teacher uses legal discourse as a strategy for deflecting Rosemary's request. By insisting upon legal documentation for oral testing, the teacher renders Rosemary's parental input as inconsequential and establishes that she alone makes decisions regarding classroom practice unless otherwise dictated by law.

Much like Marie's resistance to the school district's single educational option for her son, Rosemary resists special education services as Rebecca's only option. Rosemary recalls the professional response to her choice to implement a 504 Plan within the general education classroom rather than place Rebecca in special education.

145 R: It was fought tooth and nail from day one. Because the teachers
complained that they didn't have time to do anything different for
Rebecca or to give an oral test. They didn't have *time* to teach her
in a different way. Rebecca will sometimes learn it by the written
word, sometimes visual, sometimes auditory. It's never the *same,*
150 so that you can say that Rebecca *always* needs it this way. They
fought it tooth and nail, but I watched. I used to teach and I have
family who all teach. And I watched what was happening in
special ed and those kids are allowed—"okay they did the best they

could but they only wrote one letter today and we'll give them an A
155 [**in an ironic tone**]." They are not pushed *at all.* And I did not want her
there because you had tested her and said she had intellect, so I
pushed as hard as I could push to try to get the brain to release some
of its information. I just didn't know anything positive, plus the *social*
stigma of somebody in special ed. It is *there.* I don't care. All the books
160 and professors can say it's not, but it *is.* I didn't *want* that for Rebecca.
The more I fought it, the more they would try to dig in their heels.

In exercising her right to participate in educational decision making, Rosemary asserts that the "least restrictive environment" for Rebecca is the general education classroom with accommodations under Section 504. Rosemary describes her negotiations in terms of conflict rather than collaboration, relying twice upon the idiom "fought tooth and nail" (lines 145 and 151) to describe the degree of resistance from school professionals. She also recalls how school professionals "would try to dig in their heels" (line 161). Implicit within the general education teachers' complaints that they "didn't have time" (lines 146 and 147) to teach Rebecca "in a different way" (line 148) is the growing ideology that two types of students exist (those who belong in the general education classroom and those who belong in special education) as well as two types of teachers (general education and special education) who teach in qualitatively different ways (Reid & Valle, 2004; Varenne & McDermott, 1998). With a separate and parallel system of special education in place, the teachers do not consider it their job to educate Rebecca and openly resent being held legally responsible under Section 504 for accommodating a student who should be in special education. In other words, Rosemary insists that the teachers do something rather than merely exclude Rebecca on the basis of "not fitting."

Rosemary's narrative reflects her resistance to an authoritative discourse. For example, Rosemary uses 11 "I" sentences within 11 lines of consecutive text, grounding her opinions about special education within her own knowledge base (lines 151–153). Moreover, she mocks the low expectations that special education holds for their students (lines 153–154) and refuses "expert" claims that do not reflect her lived experience. For example, Rosemary insists that there is a social stigma associated with special education: "It is there. I don't care. All the books and professors can say it's not, but it is. I didn't want that for Rebecca" (lines 159–160).

Rosemary goes on to describe teacher resistance to a legal discourse that can mandate classroom practice.

R: They look at you like you're a fool. "Do you think we're going to
believe *those* pieces of paper? And we *can't* do this and we *can't* do
that [**in an ironic tone**]." We did not usually in the elementary school

195 have that many outsiders come in to be with us on conferences. Probably a big mistake. But, we would get some help. They would offer to do this or do that. But when we hit middle school, we had to get the school system to send someone in with us *every* time we had a conference. Because we were getting *nowhere*. They would just say,
200 "We *can't* do that. We *don't* do this." And she [the special education administrator] would have to speak up and say, "I'm sorry, but that *is* being done. That *can* be done." But the school acted like they were totally in charge and *nobody* could tell them that they had to do anything. We had to go over their heads and get somebody *from* the
205 school district to come and sit in with us.

In much the same way that Cam and Mimi describe having been perceived by school professionals as transgressing rather than operating within law (chapter three), Rosemary recalls how teachers dismiss the educational recommendations outlined in a private evaluation. Moreover, implicit within the teachers' resistance appears to be resentment toward a legal discourse that gives "outside experts" the power to determine *their* classroom practice.

Rosemary recalls such intense teacher resistance at the middle school level that it becomes necessary to "get the school system to send someone in with us *every* time we had a conference" (lines 198–199). In the face of "getting *nowhere*" (line 199) in regard to teacher cooperation, Rosemary relies upon a special education administrator to verify that her requests are indeed valid and supported by special education law. Thus, Rosemary's narrative suggests that two parallel systems of public education now exist—general education and special education—with each system vying for expression of the power inherent to their respective discursive practices. For example, Rosemary recalls that "the school acted like they were totally in charge and *nobody* could tell them that they had to do anything" (lines 203–204). In deciding "to go over their heads and get someone *from* the school district to come and sit in with us" (lines 204–205), Rosemary reflects her understanding of the tense relationship between general educators and the special education administrators responsible for enforcing special education law. Thus, Rosemary recognizes conflicts in educational decision making not only between herself and school professionals, but also between general education teachers and special education administrators.

Lindsey corroborates Rosemary's experience of teacher resistance to classroom accommodations. In a narrative that takes place within the early 1990s, Lindsey recalls an English teacher's hostility and derision toward oral testing and taped texts, as mandated on Charlie's IEP.

L: One thing I forgot to tell you about in going back to him transferring from middle school into Morgan High School was the first IEP. meeting. By then, I was pretty good at

IEP meetings. And when we came, the assistant principal was there, another teacher at random, a teacher who was going to teach his English class, who I knew by reputation since my other boys had been there; nobody had ever had her—but I knew who she was—and the psychologist who had tested Charlie. By now, they were testing his IQ at about 102, you know—whatever. I think that from time to time, they just kinda went
440 along and copied his IQ after a few years and dropped it a few to sort of fit in. If you also went back to his BSAP score, you could see every year he was learning less and less and less. This English teacher that he was supposed to have had been moved. She had been teaching gifted kids and had quite a reputation. She was mad because now she was teaching
445 average kids and Charlie's coming into this class. She said two things. This is all she said the whole time. "First is—do you expect me to give oral tests?" And I said, "No, ma'am. The resource teacher does that. And if she can't do that, then all she has to do is call Michael's mother and Michael's mother can give him the test and I'll give Michael's test.
450 I know other kids who are in this class and we can help her, but no ma'am I would not expect you to do that." And then she said, "And books on tape? Is that a joke? You will have every kid at Morgan High School coming over for that kind of recreation [**flippant, sarcastic tone**]!" And I just said, "I don't know if you have ever listened to a book, but it is so
455 hard. It is so hard, especially if you have attention problems."

Lindsey signals the introduction of a narrative by stating, "One thing I forgot to tell you about in going back to him transferring from middle school into Morgan High School was the first IEP meeting" (lines 431–433). She recalls approaching the IEP meeting with confidence, seemingly undaunted by the usual professional assemblage. After identifying the professionals present at the IEP meeting, Lindsey momentarily disrupts her narrative with an aside about testing. With a hint of derision, she remarks, "By now, they were testing his IQ about 102, you know—whatever" (lines 438–439). From the vantage point of the present, Lindsey mocks the value that school professionals attribute to IQ, implying that IQ is nothing more than a construction used by psychologists to justify their educational decisions ("I think that from time to time, they just kinda went along and copied his IQ after a few years and dropped it a few to sort of fit in" [lines 439–441]).

Upon resuming the narrative, Lindsey offers relevant contextual information about the English teacher: "She had been teaching gifted kids and had *quite* a reputation. She was mad because now she was teaching *average* kids and Charlie's coming into this class" (lines 443–445). In setting up the context, Lindsey reveals her understanding of the hierarchal arrangement of schools—high academic achievers at the top and special education students at the bottom—with teachers perceived as commensurate in status to the students they teach (Brantlinger, 1996, 1997, 2003; Brantlinger,

Majd-Jabbari, & Guskin, 1996). The introduction of a "special education" student into the class mix, a student who requires oral testing and taped texts, appears to add insult to injury for the English teacher. Like the teachers described in Rosemary's narrative, this teacher resents an educational decision that holds her responsible for accommodating a student who, in her opinion, *belongs* in special education. Furthermore, the teacher ridicules the use of taped texts to underscore how inappropriate Charlie is for her class. It is of interest, then, that Lindsey enables the teacher to resist responsibility for Charlie by relegating oral testing not only to the resource teacher, but also to herself and another resource student's mother.

Before long, it becomes clear that the English teacher intends to exclude Charlie from her class, regardless of whatever educational decisions the IEP committee makes.

L: So, anyway, the resource teacher went on jury duty that first week that Charlie was there. And Charlie went into this English class. The very first day of this class, she announced to the class—she announced to Charlie in front of the class—"You're not going to be able to keep up
460 with this group. Your reading is just not going to be able to make it in this class. Why don't you take your book in the hall and you're going to be working on this by yourself because it's not going to be fair for you to come in here and listen to the discussion on the book because these other kids have read and you can't keep up [**sarcastic tone**]." Well,
465 Charlie didn't want to tell me. The resource teacher was on jury duty, so she was gone and it was two weeks before another teacher who was a friend of mine called me and told me what was going on. And by then, you know, I was furious. I knew what she was doing was illegal, but I just said, "Get him out of there. I could spend all my energy working on her and get nowhere or I can give that energy to my child. So, just get him
470 out of there."

Lindsey sets the stage for the narrative by noting the resource teacher's absence, suggesting the relevance of this fact to the events that are to unfold. Charlie attends the general education English class mandated on his IEP as "the least restrictive environment." Despite legal grounding for this educational placement, the English teacher flagrantly resists by positioning Charlie as an outsider in her classroom—both figuratively and literally. By publicly naming his deficiencies on the first day of class, the teacher establishes Charlie's exclusionary status among his peers, followed by physically excluding Charlie from the classroom space to work in isolation from "these other kids" (lines 463–464) who presumably belong. The teacher's perception of Charlie as a static and incompetent special education learner manifests within her language ("not going to be able to keep up" [line 459], "just not going to be able to make it" [line 460], "you can't keep up" [line 464]). Inherent within her statements is an ideology that justifies segregation of students with disabilities on the basis of maintaining fairness to students *without* disabilities (lines 462–464), implying, of course, that students without disabilities deserve an education unencumbered by those students who do not belong. It is clear that this teacher conceptualizes students in relation to one another along a continuum of worth, supporting Foucault's (1975) contention that "the disciplines characterize, classify, specialize; they distribute along a scale, around a norm, hierarchize individuals in relation to one another and, if necessary, disqualify and invalidate" (p. 223). And in this case, the teacher adamantly defends and acts upon her belief that it is right and natural to disqualify and invalidate less desirable students to preserve the education of those students deemed as more valuable.

Lindsey goes on to reiterate that the resource teacher is on jury duty; therefore, two weeks pass before another teacher (a personal friend) informs Lindsey of Charlie's situation. It is striking that in the absence of the resource teacher there is no other monitoring of Charlie's education or any accountability regarding the violation of his rights, reflecting yet another example of the two separate and parallel systems of education at work. Despite her anger at this teacher, Lindsey chooses to remove Charlie from the class rather than force compliance from the teacher. Thus, Lindsey realizes that special education decisions, although *legally* mandated, do not guarantee proper implementation at the classroom level. In other words, she learns through experience that it is not possible to legislate teacher attitude.

Rosemary, on the other hand, relies upon the law to enforce educational decisions at the classroom level. In that Rebecca's father is a labor attorney, exercising rights under the law is familiar practice to this family.

R: So we finally convinced with the 504—with
the letter of the law and threatening to go to court—that she could use
a calculator in math in fifth grade. They just fought it tooth and nail.
115 And *finally* after a semester of doing that or at the end of the year, the
teacher said, "Do you know what? She didn't *cheat* on the calculator.
Rebecca *actually* did it step by step. I just couldn't believe it. She
actually doesn't use it to cheat—*just* for the multiplication tables."
There have been a few incidences. Of course, she hit sixth grade and
120 they just—*absolutely* not unless you're going to give a calculator to
every student in the classroom.

Despite having established Rebecca's legal right to a calculator in her Section 504 Plan, Rosemary recalls teacher resistance to implementing this educational decision. Rosemary, unlike Lindsey, chooses to force teacher compliance under the law. The fifth grade teacher initially believes that a calculator will give Rebecca an unfair advantage over students without disabilities. Thus, it seems that students with disabilities belong in the general education classroom *only if* they behave exactly like students without disabilities—the same ideology that justifies the segregation of students with disabilities on the basis of maintaining fairness to students without disabilities (Reid & Valle, 2004). Such ideology also emerges in the sixth grade teacher's response to Rebecca's use of a calculator ("*absolutely* not unless you're going to give a calculator to *every* student in the classroom" [lines 120–121]). Moreover, it is noteworthy that the general education teachers operate as if it is their right to refuse accommodations established under special education law—as if they are immune from regulations governing that "other" system of education.

Cosette provides yet another example of a teacher's concern about giving an unfair advantage to a student with a disability.

C: Another incident we had with that same particular teacher was after we
had my daughter tested and had been told that she was dyslexic and they
suggested that we get her private tutoring a couple of afternoons a week—
515 one thing that would be very beneficial to Lily would be to get the
classroom math book and English book so that the tutor could kind
of work with my child well before they covered that material in class.
And I approached the teacher about getting the textbook so that we could
have those at home for the private tutor and she refused to give them to
520 me and felt like it would give my daughter an advantage over the rest of
the students if indeed she had that textbook early.

Lily's teacher sends Cosette the unmistakable message that students with disabilities belong in the general education classroom only if they behave exactly like students without disabilities. Lily must perform like all the other students in spite of her reading disability. Thus, Lily is denied earlier access to textbook material on the basis of maintaining fairness to everyone. It is of particular interest that the teacher's concerns center upon the potential impact of Lily's reading disability upon "the rest of the students" (lines 520–521), rather than the impact of Lily's reading disability upon her own performance.

In the following narrative, Rosemary recalls one teacher's response to the recommendation that Rebecca receive a lecture outline to supplement her notes.

R: She was in a social studies class and one of
our recommendations was to give an outline of the lecture so that if
she didn't get all of the notes down, she could go back and see "Oh.
This was important." So, the teacher said, "I can't do that without

260 doing it for *all* my children." The end of the year, he came up to me and said, "I am *shocked* at the number of students that come up to get the outlines!" He would just lay them down. It was his *smart* students that were picking up the outlines to make sure that *they* had everything, not *just* Rebecca. That's just *not* too much to ask.

Like the teachers who appear in the previous narratives, the social studies teacher operates out of the "fairness ideology." However, he chooses to make Rebecca's accommodation available to the entire class and discovers that many students benefit besides Rebecca. In the retelling of this event, Rosemary challenges the notion that children with disabilities negatively impact the education of non-disabled children, disrupting traditional ideas about different "types" of students requiring different "types" of instruction. In a final note of resistance, Rosemary concludes, "That's just not too much to ask" (line 264).

Beyond the elementary school level, parents not only negotiate for classroom accommodations, but also for the most appropriate level of placement within general education classes. Lindsey explains the basis upon which high school teachers, in the early 1990s, made decisions about the most appropriate class placement for her son.

L: And Charlie started out doing okay, but not great, and
370 then quickly what they did for him to pass was to move him down levels in classes so that he was in a below average English class. They were not teaching to his intelligence, they were simply trying to get him the least amount of things to do.

It is noteworthy that the teachers respond to Charlie's academic struggles not by *teaching* him, but by *placing* him in a class that required "the *least* amount of things to do" (lines 372–373) in order to pass. In other words, the teachers look for a place to *put* Charlie (presumably because he "does not fit" their classes), thereby relieving themselves of the responsibility and/or burden of having to *do* something for him.

Rosemary, on the other hand, refuses to accept Rebecca's exclusion from an upper-level English class on the basis of test scores.

R: I insisted Rebecca be put in—
245 not the top English class—but the upper, and she qualified for the *very* bottom... I said, "Absolutely not. I am overriding what the tests show." They said, "Well, she *cannot* make good grades." And I said, "If she makes Cs and Ds, I will be happy because I feel like she will learn more in
250 that class than she would have in this very bottom class that has all the discipline problems." She came up making Bs and Cs! It *shocked* them to death! And *that* was what I was fighting for. Don't put her in the bottom class. I know where the—in quote—standardized discourse—in quote

255 tells us she should be. Put her in a class where *more* is expected and she *always* pulls through.

Rosemary's narrative is useful for highlighting the differing perspectives held by professionals and parents as they engage in educational decision making. For example, school professionals, relying upon their belief in the Truth of test results, recommend Rebecca for "the very bottom" (lines 245–246) English class that "she qualified for" (line 245). Rosemary, on the other hand, challenges the use of this particular Truth to justify Rebecca's exclusion. Resisting a professional discourse that "insidiously objectifies those on whom it is applied" (Foucault, 1975, p. 220), Rosemary insists that Rebecca be given the opportunity to participate in the general education curriculum. Recalling that Rebecca "came up making Bs and Cs" (line 251) and "shocked [the teachers] to death" (lines 251–252), Rosemary smugly reflects, "And that was what I was fighting for" (line 253).

Devalued Knowledge of Mothers

The narratives told thus far by this generation of mothers (mid-1980s to mid-1990s) suggest that professionals continue to assume an authoritative rather than collaborative posture toward parents. Much like the professionals described in chapter three, the professionals who appear in these narratives routinely claim scientific and bureaucratic superiority over the contextual knowledge that mothers possess.

It is of interest that these mothers, like the mothers who appear in chapter three, also mark "the beginning" of their stories not at the moment of professional diagnosis, but at the moment that they first recall having noticed something different about their children. In the first few minutes of her interview, for example, Sue shares the following:

S: Well, I guess if we start at the beginning I guess I realized that there was
5 something different about Evan. I mean, when you have your first child you aren't quite sure how they're supposed to be. And everybody in my family is weird [**light laugh**]!

Despite being a new mother (and her wry observation that "everyone in my family is weird" [lines 6–7]), Sue remembers having had an intuitive feeling that "there was something different" (lines 4–5) about her firstborn.

As an experienced mother, Dawn, unlike Sue, is able to make comparisons between her two older children and her third child, Brittany.

D: But then when it came along with Brittany, she's three years younger than Shawn, as I said before, I noticed when she

> 20 was six months old that she was a little bit slower than all the rest of them. I didn't realize that she would be dyslexic. I didn't realize that she would have a problem. But I *knew* she was different as far as her ability to catch on to things. She didn't walk as fast. Another very interesting mother's observation is she had a difficult time with wetting the bed at night. I
> 25 found out that there is some kind of neurological reason there and that also prevents them from learning. And that definitely happened.

Echoing the narratives told by Della, Lindsey, Cosette, and Mimi (chapter three) in which they identify infancy and toddlerhood as the starting point of their uneasiness, Dawn traces her first suspicion back to Brittany's infancy. From the vantage point of the present, Dawn sees the connection between her early observations and Brittany's eventual diagnosis of dyslexia. It is noteworthy that Dawn introduces the fact of Brittany's bedwetting by stating, "Another very interesting mother's observation is" (lines 23–24), thereby distinguishing the knowledge that mothers possess as particular and contextual. Yet, Dawn appears to substantiate the validity of her own observations by offering professional evidence for "some kind of neurological reason" (line 25) that connects bedwetting to learning problems. Concluding that this connection "definitely happened" (line 26), Dawn appears satisfied that her concerns about Brittany's bedwetting were justified.

Chloe, mother of three children, recalls having had nagging concerns about her youngest child's struggles in kindergarten.

> C: So we went ahead and we repeated kindergarten. She's one of the few kids I know whose *mother*
> 75 kept them back in kindergarten, *but* I thought it was best for her. So when we got *halfway* through kindergarten, she was still doing the same thing. She was *way* above the other kids in most of the subjects, most of the learning, what they were supposed to be doing, but when it came to the *writing* and it came to *reading* anything—she just couldn't do it. She
> 80 could converse with you, she could memorize at the drop of a hat, but she couldn't write anything and she most certainly couldn't read. Right towards the end of her kindergarten year, I decided to have her tested because I *knew* that this was something that was running in our family now—that we were having children with learning disabilities.
> 85 My son had had it. I figured Ashlyn might. So, we arranged to have her tested.

In the same way that the first generation of mothers recognizes developmentally different behaviors in their young children, Chloe is well aware that her daughter has not yet mastered the kindergarten curriculum. It is noteworthy that Chloe, not the teacher, suggests that Ashlyn might benefit from a second year in kindergarten. Halfway through the second year of kindergarten, Chloe observes Ashlyn "still doing the same thing" (line 76), despite her second exposure to the curriculum. On the basis of her own observations, Chloe concludes that Ashlyn demonstrates considerable strengths in verbal expression, reasoning skills, and memory, yet "she couldn't write anything and she most certainly couldn't read" (line 81). Drawing upon her mother knowledge and experience with her older son, Chloe decides to have her evaluated in order to confirm or rule out a learning disability. It is of interest that Chloe and the kindergarten teacher do not appear to collaborate in any meaningful way regarding Ashlyn's struggles in reading and writing. In fact, Chloe appears to take the lone initiative in thinking about and responding to her child's academic struggles.

As mother to four children, Rosemary recognizes early on that Rebecca learns in a qualitatively different way than her three older and academically successful siblings.

R: I had three kids that were honor students. You just sort of put it in front of them and they just absorbed it and made honor roll. Here comes Rebecca. And I'm doing all this other stuff with *them*—just sort of pushing it out to them and they absorb it. And then there's
65 Rebecca. You go over it and over it and you ask her the question again and she has no idea what you've just been studying for thirty minutes. And saying it over and over and over again. In the next five minutes, she might tell you *exactly* what you said. But you'd come back and say, "Now what was that?" And she couldn't tell us. It was a
70 very *frustrating* thing.

Using her other children as a point of comparison to Rebecca, Rosemary describes how she "just sort of put it in front them and they absorbed it and made honor roll" (lines 61–62), implying a "taken-for-granted" family expectation for school success. Rosemary foreshadows the subsequent disruption of this expectation in her segue, "And then there's Rebecca" (lines 64–65). In working closely with Rebecca, Rosemary learns that "you go over it and over it and you ask her the question again and she has no idea what you've just been studying for thirty minutes" (lines 65–67). Although she does not yet know why Rebecca struggles to learn, Rosemary closely observes how she responds to instruction. For example, Rosemary recognizes that Rebecca inconsistently retains and retrieves learned information. Drawing upon her contextual knowledge of Rebecca, Rosemary already understands what professionals would later name as a receptive and expressive language disorder.

Despite such evidence of the rich contextual knowledge that mothers have regarding their children's language, behavior, and abilities, they continue to report instances in which their knowledge is disregarded or dismissed by professionals. In the following narrative set in the mid-1990s, Chloe describes how professionals respond to her concerns about Jessie's declining performance in mathematics.

C: We didn't see the learning disability. Things I was seeing in Ashlyn, I wasn't recognizing in Jessie. And I should have. When her eighth grade year came up, I went in for the
295 conference with teachers—which I did twice a year like you're supposed to and nothing had ever come up before. Towards the end of the eighth grade year when they were scheduling them for high school, she was put in all these advanced courses. But I was seeing math as a steady decline when she got from just the basics and started coming into the
300 pre-algebra. I was seeing her grades starting to decline. Where she had been good strong Bs, she was falling to just B. Then it was a B minus. Then it was C plusses. Then it was Cs and C minuses. But the math teacher over there said, "You know what—this is what happens in middle school. Kids just get, you know—their hormones go crazy, they're going
305 through puberty, they're going through all this junk and all of this, you know, peer pressure and kids and 'Your clothes stink' and this and that, you know, 'You're not dressing the way we are and your mother doesn't drive this kind of car'—all these stupid little things that come in." So they said, "She's probably *fine*." But I was looking at math papers and I was
310 seeing things that were really bothering me. But it was just—"No, she's just going through a rough period."

Chloe begins the narrative with an abstract (Labov & Weletzky, 1967) that summarizes the point of the story she is about to tell: "We didn't see the learning disability. Things I was seeing in Ashlyn, I wasn't recognizing in Jessie. And I should have" (lines 289–291). In much the same way that she describes her lone initiative to understand Ashlyn's academic struggles in kindergarten, Chloe takes sole responsibility for not having noticed that her middle child, Jessie, showed "the signs" of a learning disability, as well. Thus, Chloe appears to understand that her role as the mother is to persuade rather than to collaborate with school professionals. Her lived experience thus far leads her to believe that parents alone bear responsibility for recognizing their children's learning problems.

Having established the orientation of the narrative (Labov & Welletsky, 1967) within the context of a parent/teacher conference, Chloe implies that she is a cooperative and involved parent because she regularly attends such conferences "twice a year like you're supposed to" (lines 295–296). By noting that "nothing had ever come up before" (line 296), Chloe suggests that routine parent/teacher conferences primarily function as a vehicle through which teachers inform parents. For example, Chloe states that "they" (line 297)—presumably the teachers—inform her that Jessie has been "put into all these advanced courses" (line 298) for high school. Although it is not clear that "they" invite her to respond, Chloe recalls expressing concern about Jessie's placement in an advanced math course, pointing out the "steady decline" (line 299) in her math grades. The math teacher immediately discounts Chloe's concerns, choosing instead to authoritatively expound upon "what happens in middle school" (lines 303–304) rather than consider a mother's particular concerns about her particular child.

Chloe uses the conjunction "but" four times within this narrative, signaling resistance to professionals who claim authority over her own knowledge. For example, Chloe makes the following assertions: "But I was seeing math as a steady decline" (lines 298–299) and "but I was looking at math papers and I was seeing things that really bothered me" (lines 309–310). However, the professionals respond with such patronizing statements as "She's probably fine" (line 309) and "no, she's just going through a tough period" (lines 310–311), revealing their disregard for "the kind of knowledge women value and schools do not" (Belenky, Clinchy, Goldberger, & Tarule, 1997, p. 313). Rosemary recalls the professional response to her insistence upon the general classroom as the "least restrictive environment" for Rebecca.

> R: They told me that it was my pride keeping Rebecca in the regular classroom. It had nothing to do with Rebecca. It was my pride. That
> 185 I didn't want her in special ed or special services. Oh, I was angry, furious! I just knew that was not the case for her. I'm not dumb!

By rejecting the school recommendation for special education placement, Rosemary disrupts the conventional power distribution between parents and professionals. In response to having had their authority rebuffed, the professionals, in turn, interpret Rosemary's resistance as evidence of a mother in denial. There appears to be no acknowledgment whatsoever of Rosemary's knowledge about her own child. She recalls her indignation at the assumptions made by professionals and insists upon the validity of her knowledge. In much the same way that Cam (chapter three) reassures herself in the face of professional judgment ("it's okay, Cam. You're *not* an *idiot*"), Rosemary emphatically asserts, "I'm not *dumb*!" (line 186).

Influence of Race/Culture, Social Class, Gender

Like the mothers in the previous generational group, these mothers are White. All live in the southeastern United States, except for Marie who resides in a northeastern state. Three mothers (Rosemary, Chloe, Sue) live in a medium-sized city. Marie resides in a small suburban town. Dawn represents the only participant living in a rural context.

None of the narratives directly addresses the influence of race/culture upon interactions between parents and professionals. This is not particularly surprising given that all of the mothers are members of the dominant culture. Moreover, four of the five mothers live in the southeastern United States where, in some contexts, it remains uncommon to speak aloud about race.

Two mothers, however, allude to the intersection between race and public education. For example, Dawn makes oblique references to race while reflecting upon reasons that some parents resist having their children labeled as "special education" students.

335 D: Then you have the probably very uneducated parent who never finished school, who never had an opportunity to know anything—especially here in the South—they probably would not accept a lot because they are afraid to. They probably never learned to read. So to them, they did what I did for a long time—refuse to read. I read because of what I had to read, but I
340 sure never read for pleasure. I never read because I really wanted to read. I was forced to read. I read what I had to read to get by. I think a lot of very—I'm trying to say it with the right words—low level or low educated families, it's probably the way they have to deal with it. Going back to what I said before, a little education would go a long way with these
345 parents.

As an insider to Southern culture, I understand that the parents to whom Dawn refers, but does not name, are African American. It is not uncommon for White Southerners to rely upon euphemisms when discussing race. By adding "especially here in the South," Dawn communicates that she is, indeed, referring to race—an insider code that I, as a White Southerner, immediately recognize. By self-monitoring her word choice, Dawn again signals the use of euphemistic language for race: "I think a lot of very—I'm trying to say it with the right words—low level or low educated families" (lines 342–343).

Dawn appears to believe that African American parents are encumbered by low ability and/or education; thus, she concludes that African American parents resist special education services because they are uninformed. Although she acknowledges her own reading problems, Dawn seems to differentiate herself from rather than identify with "these parents" (lines 344–345), suggesting that "a little education would go a long way" (line 334) toward resolving their issues. Dawn's focus upon perceived deficiencies in African American parents (rather than upon the social, economic, and political context in which they live) reflects her positionality within the dominant culture. Given that the American teaching force draws primarily from the dominant culture, it is relevant to consider the impact that a middle-class standpoint might have upon parent/professional relationships (Brantlinger, 2003, 2004; Habermas, 1978; Tyack & Tobin, 1994).

Chloe, drawing upon her experience as a former PTA president, comments upon the sense of entitlement that she observes among White parents.

C: And they are all about the gifted and the rest of the kids— the normal and the kids with learning disabilities—and they are like "Why are they in school? Why don't their parents have them home picking cotton [**in**
1215 **a sarcastic tone**]?" Well be to the teachers who have these parents. When their kids are in their rooms, they better pass. We had people on the PTA that were after the two guidance counselors here because they had the guts to stand up for some people and they wanted them fired. Went to the district and tried to have them fired [**pause**]. One of them was
1220 Black [**voice lowers**].

Chloe's language belies her discomfort in speaking about race. For example, she does not use the term "White parents," instead referring to "they" (lines 1212 and 1213) and "these parents" (line 1215). In retelling the incident in which "people on the PTA" (lines 1216–1217)—a euphemism for White parents—want two guidance counselors fired, Chloe pauses, as if hesitant to transgress, then lowers her voice to reveal that one of the guidance counselors is "Black" (line 1220). The racial implications remain unstated because she assumes that I, a White Southerner, can fill in the blanks—as I do.

Chloe's observations reflect multiple points of intersection between race and public education. In stating that White parents "are all about the gifted" (line 1213), Chloe implies that there is an educational hierarchy that runs along racial lines—a hierarchy in which White parents heavily invest themselves—with White children primarily populating gifted classes while "the rest of the kids" (line 1213), including "normal and the kids with learning disabilities" (line 1214), remain segregated in lower levels of the hierarchy (Brantlinger, 2003). Having observed the way some White parents disregard "other" children, Chloe mocks their attitudes of entitlement and exclusion—supporting Brantlinger's assertion that educated middle-class parents "do not think beyond their own children when they interact with schools" (p. x). In her use of hyperbole, Chloe communicates not only how significant she believes this exclusionary stance to be, but also how closely related it is to the history of racial oppression in the South—specifically reflected in her reference to "picking cotton" (line 1214).

Chloe further comments upon the ways in which White parents exercise the power afforded to them as members of the dominant culture. On the basis of Chloe's insider perspective as PTA president, it appears that a parent's racial and/or cultural positionality influences how effectively he or she may be able to negotiate within the school context. Thus, it might be assumed that the White mothers who appear in this chapter, by virtue of their membership in the dominant culture, possess an advantage, whether they recognize it or not (Bell, 1994; hooks, 1989; Ladson-Billings & Tate, 1995), that mothers outside the dominant culture are unable to access.

Yet membership in the dominant culture, however advantageous, does not guarantee an equitable place for mothers in the decision-making arena. Much like the preceding generation of mothers, these middle- to upper-class White mothers rely upon material resources available to them as a means for enhancing their negotiating power. For example, Rosemary recalls resisting professional domination by strategically drawing upon the social power afforded to her husband as a lawyer.

500 R: We were not *rich* rich. But we were above middle class. We didn't bring
a lawyer because John is a lawyer. He always had to wear his "lawyer
suit." They thought that John with his dark suit and tie with his arms
folded might have had power. He's a labor lawyer. But, he didn't have any
power in town! They don't even deal with town stuff, but they didn't *know*
515 that. He'd have to go hire somebody [**light laughter**]! He acted ugly at
times. And it used to just embarrass me to death. It did. He would have to
threaten, you know. And it was awful. Sometimes to get *anything* done,
you had to threaten.

By referencing her socioeconomic status, Rosemary reveals awareness of how social class may contribute to her negotiations. In other words, she expects that the privileges typically afforded to her will be similarly advantageous in her negotiations with school professionals. It appears, then, that parents of higher social status are better positioned to influence school professionals than parents of lower social status (Apple, 1992; Oakes, Gamoran & Page, 1992; Tooley, 1999). For example, Rosemary points out that John, a lawyer dressed in a "lawyer suit" (lines 501–502), commands the power granted to him by society, a power to which he expects school professionals to respond—despite the fact that, as a labor lawyer, he knows little about special education law. Rosemary recalls how John strategically relies upon his cultural and social status to intimidate and pressure school professionals.

Despite the promise of a "free and appropriate public education" for all children (IDEA, 1997, 1990, 2004), these mothers report continued reliance upon private resources (e.g., assessments, schools, tutors). For example, Cosette outlines the private resources that contributed to Lily's educational success.

345 C: We had her tutored for two years in the public school system with private tutors and then we entered her into Crossroads Academy which is a school designated for children with special needs and learning disabilities. She attended there for two years and did quite well and went into a private school for the remainder of her schooling years and actually
350 graduated with honors from Green Valley, she did. And then went on to a private college and graduated with a degree in sports medicine.

In negotiating Lily's school career, Cosette accesses the private resources available to her. Lily benefits from private tutors, private school (special and general), and private college; yet, Cosette, an affluent member of society, does not acknowledge the benefit of financial assets (Conley, 1999; Oliver & Shapiro, 1995) nor does she appear aware of the class privilege inherent within Lily's particular experience of disability.

Rosemary likewise depends upon private tutors to supplement Rebecca's public school education.

R: We studied for tests. That's when I lost it—*really*, being a mother—you *cannot* do both. And it's *so* expensive. If somebody doesn't have the money, I don't know how you would do it.
340 It was *very* expensive to send her to all this tutoring.

In an effort to maintain a healthy mother/daughter relationship, Rosemary relinquishes her role as teacher and turns to private tutoring (line 338), a service she describes as "very expensive" (line 340). In contrast to Cosette, however, Rosemary acknowledges the role that class privilege plays in her daughter's education, reflecting upon the circumstances of families without the resources to fund private tutoring.

Dawn, who self-identifies as lower-middle class, does manage, however, to pay for private tutoring to supplement her son's "special" education. In fact, she recommends that other parents should "get the tutor after school." Thus, it appears that mothers of this generation continue to supplement their children's public education with private tutoring, much like Della reports doing during the 1960s before the passage of PL 94–142 (see chapter three).

Chloe represents the only mother who challenges the necessity of private services in light of the guarantee of a free and appropriate public education for all children (IDEA, 1990, 1997, 2004). In the following excerpt, Chloe explains her decision to return her daughter to public school after several years at a private school for students with learning disabilities.

C: It was
160 wonderful. It was a very expensive school. And it got to the point where

I would have *loved* to have kept her there, and yet, I also am a firm believer that in this country we have our public schools and our public schools should come up to the *same* academic excellence that a private school that costs $20,000 a year does... And I wanted my child in
175 public schools because I felt that she needed that exposure to other children that weren't *all* privileged children.

Although her child benefits from the class privilege that makes attendance possible at a private school for students labeled LD, Chloe chooses to return Ashlyn to public school on the basis of her belief that "our public schools should come up to the *same* academic excellence that a private school that costs $20,000 a year does" (lines 162–163). Moreover, Chloe appears to understand the intersection between social class and disability.

Echoing the narratives told by the previous generation, these mothers confirm that gender issues continue to influence interactions with school professionals. The narratives told by this generation suggest that school professionals, who are predominantly women, continue to hold traditional and patronizing attitudes toward mothers. For example, all of the mothers recall that school professionals viewed them (never the fathers) as solely responsible for their children's educational well-being. It is of interest, then, that school professionals afford such little respect to those perceived as bearing the brunt of responsibility. For example, Rosemary recounts how differently school professionals respond to her in comparison with her husband.

215 R: They'd just sort of nod their heads and pat you on the back and say, "We're going to do it *our* way. I'm sorry. You're wrong." I am not as forceful as John. I *know* I'm not. I started asking him to *please* go. I'd come home from meetings with teachers and say, "John, they are not going to change anything. I have been and done this. You go." Yes, *every*
220 time because they didn't care. I could just preach and preach and preach and get my back to the wall. He would go in his lawyer suit and sit there with his arms folded and *dare* them. And it *did* intimidate. They did what *he* said.

It is noteworthy that school professionals (identified as women in an earlier section of the narrative) take on traditional male posturing in their response to Rosemary. For example, they display patronizing behaviors ("they'd just sort of nod their heads and pat you on the back" [line 215]), dominance ("we're going to do it our way" [line 216]), and superiority ("I'm sorry. You're wrong" [line 216]), illustrating Shildrick's (1997) assertion that women who claim sameness with masculine subjectivity reinscribe rational and objective "ways of knowing" as ideal and normative.

Rather than challenge those who position her as passive, Rosemary appears to blame herself for failing to be aggressive enough. Recognizing the power inherent within John's male privilege, Rosemary solicits her husband's presence at subsequent meetings. In contrast to the manner in which Rosemary becomes dismissed by school professionals, John demands respect through intentional intimidation. In the presence of a "bonafide male," the female professionals appear less confident in their male posturing and assume a more submissive stance.

As a professional in the field and the mother of a child labeled LD, Marie is cognizant of ways that both roles play out within the context of special education meetings. She recalls consciously choosing to perform the role of professional rather than the role of mother.

M: But remember, I was a *professional* in the field who
simultaneously, as fast as I was learning and helping my own child, I
was helping my clients and I was becoming known in the community
as a professional. And I was on the speaker's circuit. So these teachers
625 would be at workshops or conferences and I would be the speaker. So
they were getting to know me. So when I came into a room, they *never*
saw me—from very early on, maybe during the early intervention years,
but from kindergarten on—they were now knowing me as a professional.
And *that's* who they saw walking into the room before they saw a mother.
630 And I learned very early on to separate my motherness out and to
present myself as a professional. I felt, I think, looking back—and I've
known it for a long time because I have lots of friends who have been
through these meetings and all my clients—I certainly had the advantage
being in the field because they never responded to me as a mother.

In light of her professional presence in the community, Marie carries her expert status into special education meetings. It is worth noting that Marie appears unable to conceptualize a space in which she can be mother *and* professional; rather she must choose one or the other, so she opts to perform the role that affords her the most power.

Marie's understanding that she must act like *either* a professional *or* a mother, but not *both/and*, reflects an unnatural dichotomy of identity roles (i.e., mother vs. professional) within the context of special education meetings. For example, Marie recalls being perceived only as an expert by teachers (most of whom are women and mothers but perform only the role of professional) familiar with her professional work. However, Marie unwittingly participates in her own identity fragmentation by consciously performing one identity at the expense of the other. As an insider to the profession, Marie recognizes the weak position of mothers and strategizes accordingly.

It is clear that Marie's professional persona enhances her status within special education meetings. In contrast to what she regards as her earned status, Marie reflects upon the status granted to her husband by virtue of his gender.

M: I think we were seen as equals initially which used to bother me because he knew *nothing* about it. He read very little. He was *not* learning and he was treated as *my* equal in this arena because he was a man in a suit. I would come in as the professional
660 and they *knew* it. There was always that initial chit-chat and "Oh, I heard you speak here and there" or "I see you are on *this* program" or "So and so heard you speak" and blah, blah, blah. They would definitely, speaking to me, have a real respect. I saw them afford that same respect to my husband and that was *fine.* I didn't even really get it initially... I would
670 come and present a very professional argument and here are the scientific facts and showed them all their first picture of the PET scan of the brain and all of that. Nobody had ever seen that before. I got it right hot off the presses when Alan Zametkin presented his first study. I was there. So *then* when they would turn to him and say, "Well, John, what do you think?" I
675 was a little *perturbed* [**laughter**] to say the least. Because what *could* he think? He hadn't done his homework [**laughter**]! But fortunately, he learned to say, "What *she* said [**laughter**]."

In making her point about the gender bias inherent within such parent/professional interactions ("[John] was treated as *my* equal in this arena because he was a man in a suit" [lines 658–659]), it is of interest that Marie reinscribes the parent/professional dichotomy between herself (professional) and her husband (parent). For example, Marie draws upon scientific knowledge to assert her credibility in order to prove her superiority over John's knowledge. Using her professional knowledge to challenge the privilege of male embodiment, Marie unwittingly discounts the knowledge that John brings as a parent, revealing her own engagement in professional dominance.

After relocating to a different area of the state, Marie compares her former interactions with school professionals to her experiences with new school professionals unfamiliar with her professional status.

M: But now these were new people who didn't know me and they didn't know me professionally. It was a *whole*
850 new ballgame. So now we walk in and, *yes*, John is *the* man. "Hello, Mr. Strauss. Can I get you coffee?" All these *women* who are administrators and teachers and psychologists and this and that acting like they are *gofers* for my husband. And I made a conscious decision to let me see
855 if I can be professional *mom*... And it was very different. I didn't get the respect. I got *questioning* of *every*thing I said and a lot of placating.

> "And I'm sure that *you* think that he won't do well in this situation" or "*you* think that he needs a laptop" or "*you* think he needs a special
> 860 program, but *we'll* go and see how he is this year or this quarter and then *we'll* decide." Of course, you had to go through procedure anyway and I understood that. But it really was the *intonation* that was dripping off the words that was important—that was *all* around it.

As a result of making a conscious decision to perform only the role of "professional *mom*" (line 854), Marie recalls experiencing "a *whole* new ballgame" (line 850) within the context of special education meetings. She observes school professionals deferring to John because of his male embodiment. Without the credibility of her professional status, Marie reports that she "didn't get the respect" (lines 855–856) and, like Rosemary, she experiences school professionals as superior ("I got *questioning* of *every*thing I said" [line 856]), dominant ("*we'll* go . . . and then *we'll* decide" [lines 860–861]), and patronizing ("but it really was the *intonation* that was dripping off the words that was important" [lines 862–863]).

Conclusion

During this historical time frame (mid-1980s to mid-1990s), the mandates required by PL 94–142 establish special education as *the* socially agreed upon institutional response toward educating American children with disabilities. Professionals adopt the "discourse of special education," a *particular* way of talking about and responding to struggling learners, as the taken-for-granted mode of communication for speaking about children with disabilities in public schools. However, as mothers of this generation attest, integration of a new discourse, even one mandated by law, is neither a simple or swift process. One to two decades after the law's passage, mothers continue to report confusion about the language professionals use to describe their children. Moreover, participation in educational decision-making requires that mothers be able to negotiate a "discourse of law," a "discourse of science," *and* a "discourse of institutional implementation."

This generation of mothers begins to recognize how special education practice, with its emphasis upon the individual as the unit-of-analysis, operates in a way that constructs children as the sum of their deficits—a conceptualization often in conflict with mothers' contextual understanding of their children. Additionally, the institutionalization of special education fosters a growing ideology that two types of students exist—those who belong in general education classrooms and those who belong in special education—the latter becoming increasingly regarded by general education teachers as falling outside their responsibilities. Thus, mothers of this generation report persisting tensions between parents and professionals in regard to shared decision making and IEP implementation, and, like the mothers of the previous generation, draw upon their own cultural collateral (social class, material resources, and male power) to gain their rightful access to the process.

Chapter Five

The Maintenance Years: Third Generation Mothers

I am defining the third historical time frame as the years between the mid-1990s and the present in which special education, a public school institution now for more than 20 years, is the naturalized educational response to students who demonstrate difficulty learning. I introduce five mothers whose children, born either in the late 1980s or 1990, are currently of school age and labeled learning disabled (LD): Elsie whose daughter Emily was born in 1988; Kim and Kanene whose respective sons, Jacob and Marvin, were born in 1989; Katie whose son Louis was born in 1990; and Debra whose son, Justin, was also born in 1990 (see appendix B). I also include Chloe (introduced in chapter four) whose later narratives cross into this time frame (see appendix E).

Kim, Katie, and Debra are White; however, Kim and Debra define their cultural backgrounds as Appalachian and Italian American, respectively. Elsie identifies as Hispanic with Puerto Rican heritage. Kanene is African American. Three mothers (Kim, Katie, and Debra) identify as middle class and reside in small- to medium-sized cities in the southeastern United States. Two mothers (Elsie and Kanene) live in a large metropolitan city in the northeastern United States and live at or below poverty level (see appendix C).

Two participants chose pseudonyms for themselves and their family members. The remaining three mothers, like the mothers in chapter four, rejected the use of pseudonyms, electing instead to reveal their identities. These mothers likewise describe their "identity disclosure" as a political act—an opportunity to name their experiences aloud.

Participant Snapshots

Following the format of chapters three and four, I begin this chapter with participant snapshots, or brief biographical sketches, to acknowledge each mother as an individual with a particular story to tell as well as to orient the reader to the general positionality of each participant. I also describe the nature of my relationship with each mother.

Elsie

During my first semester as a doctoral student, I enrolled in a graduate seminar in which students assisted professors on research projects in the public schools. In light of my special education background, I was assigned to a research group studying inclusion in an elementary school. One aspect of the project involved a collaborative of university representatives and public school stakeholders who convened monthly to discuss the school's progress toward meaningful inclusive practices. As a member of this team, I met and subsequently worked with Elsie, the project's parent representative. Our collaboration on this project marked the beginning of an ongoing professional and personal relationship, including coauthorship of a journal article (see Valle & Aponte, 2002) about parent/professional collaboration under the Individuals with Disabilities Education Act (IDEA).

Elsie begins her narrative at the moment Emily, a six-week-old African American foster child, comes into her life. In response to a social worker's request to consider taking a cocaine-positive infant experiencing convulsions and withdrawal symptoms, Elsie responds without hesitation: "Yeah, yeah! Just bring her!" She recalls, "So they brought her. My family was there. We were *all* waiting for her!" And Emily becomes enveloped by the family that is to become her own.

As the mother of a cocaine-exposed infant in the late 1980s, Elsie reads what little literature is available at the time and becomes a participant in a research study about the cognitive and behavioral outcomes of "crack babies." Yet, she relies far more upon her own observations of Emily as well as her "mother knowledge" (gleaned from raising two biological children and other foster children) than any professional resource. Elsie remembers, "Since I was experimenting really and doing all these things that were working with Emily, the professionals were taking that and giving it to other parents to use at the time."

With the advent of Emily's preschool diagnosis of a speech/language disorder, Elsie begins a long-term relationship with special education professionals that continues to date. By educating herself about special education law ("I had no one else but *books*! You know? Nothing else but books!"), Elsie eventually gains the negotiation skills necessary to fully participate in educational decision making. However, she quickly learns that successful negotiation for services does not necessarily translate into quality education in the classroom. Recognizing the limitations of the elementary inclusion classroom Emily attends, Elsie becomes a full-time classroom volunteer to help the teacher meet the needs of *all* students. Inspired by her experience in the classroom, she subsequently commits to being a full-time volunteer throughout Emily's elementary school years.

Elsie reflects, "Everybody knows by this time, my focus is special ed, ADD, learning disabilities, all this stuff. I'm looking at it from this angle." Considering herself an advocate not only for her own child but for all children, Elsie regularly offers her hard-won knowledge of special education to other mothers. With almost missionary-like zeal, Elsie capitalizes upon any opportunity to raise public awareness about learning disabilities. She admits, "It is my life. I have to say that it is my life."

Kim

In 1990, Kim realized her dream of motherhood with the adoption of a newborn, Jacob. Within the next five years, three siblings (local and international adoptions) join the growing family. Fourteen-year-old Kostya, a recent arrival from a Russian orphanage for physically disabled children, makes the fifth sibling. In addition to managing her lively household, Kim conducts home studies for prospective parents seeking adoptions. I am honored to be Jacob's godmother and a longtime friend of this vibrant family.

Kim's narrative opens as she observes four-year-old Jacob in his Vacation Bible School class. She recalls, "He's gazing up at the teachers with *stars* in his eyes, on the edge of his chair, just taking in every word they were teaching. And so, I thought, 'Oh, what a perfect little student he's going to be!'" Jacob's subsequent academic challenges and eventual diagnosis of a learning disability take Kim down a path unanticipated.

Kim's story chronicles how school personnel shift perception of her as the dedicated and welcomed "volunteer mom" to the tiresome and demanding "special education parent." After struggling with school personnel throughout Jacob's elementary school years, Kim opts for a private school for students with learning disabilities during his middle school years—an experience she describes as "everything you had ever wanted in an IEP without *asking* for it!" With Jacob's educational needs met at the private school, Kim sees herself become a "more well-rounded mom" because of her increased availability to *all* of her children. She notes, "It has been the best thing for my *whole* family."

Reflecting upon Jacob's educational journey thus far, Kim comments, "I feel like I'm so glad I've walked this road because I am so much stronger that I would have been. And I'm in a position that I can truly be what Jacob needs me to be—not the one *pushing* him, but the one standing beside him and celebrating the differences he has. I'm just grateful for the experience and I know it's made me a better mom."

Kanene

Unlike most of the mothers in this study with whom I have had long-term relationships, Kanene and I met for the first time on the day of the interview. Her son's special education teacher, a former graduate student of mine, referred her to me as a potential participant. Upon learning more about the nature of my study, Kanene immediately agreed to participate and arranged a date for the interview.

On the day of the interview, I meet Kanene at her son's large urban high school just after the last class ends. Having been granted prior administrative approval, we conduct the interview in the guidance counselor's office. Kanene asks about my professional background and, in particular, my interest in parent/professional relationships. Following a few minutes of conversation, she expresses eagerness to share her experiences and appreciation for the opportunity to do so. We settle comfortably into the interview.

Kanene's narrative traces her evolution from an uninformed parent to a passionately engaged advocate for her son Marvin. Her introduction to special education begins with Marvin's kindergarten diagnosis of a speech/language disorder. Frustrated by the lack of accessible information about special education options ("they don't guide you. They don't tell you"), Kanene uses the school district's document library to educate herself about IDEA and her parental rights. She learns, however, that knowledge alone does not guarantee appropriate services. Experience teaches her that "you gotta go and keep fighting and fighting and fighting"—a commitment that requires both time and energy. Kanene, a working single mother raising two boys, explains that it is not easy to take time off from work to engage in school district negotiations. But she concedes, "You got to keep on. You can't let them tell you, 'Well, we can't do this.'" Reflecting upon both the successes and challenges of educating her child, Kanene muses, "Well, you take it one day at a time, you know. You just ask God to guide you the best way He can."

Katie

Katie and I share a friendship that spans more than two decades. Having begun our friendship as young adults, we have seen one another through the predictable stages of adulthood—dating, marriage, and childbearing. Katie is a stay-at-home mother of four active sons. Over the years, our families have enjoyed a close and ongoing relationship.

Throughout her narrative, Katie returns again and again to the exclusion and stigma associated with her oldest son's learning disability. Her story begins with Louis's entrance into formal schooling—the private Catholic school that Katie and all of her siblings attended. In response to his eventual diagnosis of a learning disability in second grade, Louis's classroom teacher claims that her time cannot be compromised by one child's needs and recommends private tutoring.

During his elementary years, Louis fends off taunts from peers ("you're just *so* stupid," "you *can't* read," "you're one of the *worst* readers!") while simultaneously working "*nonstop* to catch up, but never *fast* enough to meet the demands that the school was asking of him." With a subsequent transfer to public school and access to special education services, Katie hopes that Louis can finally "have his childhood back." However, Louis's peers respond derisively to his status as a resource student: "Oh, you're in one of those *retard* classes!" Katie further notes that stigma and exclusion extend into her own relationships with other mothers who do not have children with disabilities. She explains, "It's almost like saying that my child has a fatal disease. And they go, 'Ooooh.' It's a little pity sound in their voices. And they quick try to change the subject."

Reflecting upon the stigma and exclusion her child has experienced over the years, Katie muses, "I would love to see a program initiated within the school system to educate *all* children on what the resource program does, who these kids are...then they might understand. Hopefully, it would negate the insulting that goes on due to misconception."

Debra

Debra is the youngest sibling of five, and only daughter, in my husband's close-knit family of origin. In light of our family relationship, Debra sought my professional opinion regarding Justin's preschool diagnoses of dyspraxia and a speech/language disorder as well as his diagnosis of a learning disability in second grade. Over the years, we have engaged in many conversations about Justin's education.

Debra's narrative chronicles her growth from passive parent participant to confident advocate. Upon first learning of Justin's learning disability, Debra places her faith in "the experts" to do what is best for her child ("we just sat there and listened to them and signed off on anything they told us to sign off on because this was so new to us"); however, she soon comes to realize that an individual education plan (IEP) does not *guarantee* proper implementation at the classroom level. Debra traces her yearly struggles with classroom teachers, including misperceptions ("they thought Justin was *lazy* and we as parents needed to just talk with him"), lack of awareness ("the teacher had *no idea* that he had an IEP"), and lack of accessibility to the general education curriculum ("his teachers expected him to keep up with all of the other kids and didn't realize there was anything different").

Debra's experiences teach her to assume an advocacy stance. She realizes, "As his parent, *I'm* his voice. *He* doesn't have a voice." In shifting from a passive to proactive position, Debra recognizes that her voice must be heard to keep Justin from "slipping through the cracks." She worries about other parents who do not speak up or do not know how to speak up. She reflects, "You want to be heard for *other* kids... you begin saying, 'You know what? I have to fight for everybody who isn't fighting.' As silly as that may sound, it starts to take on a whole new meaning."

Third Generation Mothers: The Collective Narrative

In the following section, I present a collective narrative crafted from the mothers' individual stories that take place between the mid-1990s and the present. The experiences reported by these mothers reveal their perceptions of the power distribution within special education discourse, how they position themselves within and against special education discourse, and the consequences of special education discourse in their lives. The stories are grouped into four broad themes: (1) the language of experts; (2) conflicts in shared decision making and IEP implementation; (3) devalued knowledge of mothers; and (4) influence of race/culture, social class, and gender.

The Language of Experts

Two decades after the passage of the Education for All Handicapped Children Act (PL 94–142), the institution of special education is well established within the public school system. For example, during the 2001–2002 school year, 2.9 million students, ages 3 to 21, received learning disability services—more than triple the number of students identified and served during the 1976–1977 school year. This indicates that the number of special education classrooms and special education professionals have increased in response to the burgeoning population of students identified as LD.

By the 1990s, a growing number of critics question the efficacy and ethics of what has become, in essence, a parallel system of education. Inclusion proponents offer an alternative "disability discourse"—with a focus on *inclusion*, rather than exclusion, of students with disabilities (e.g., Allan, 1999; Ballard & McDonald, 1999; Brantlinger, 1997; Lipsky & Gartner, 1996, 1997; Sapon-Shevin, 1999). In response to evidence of special education students being routinely excluded from the general education curriculum, the law's 1997 reauthorization mandates that special education students have access to the general education curriculum (IDEA, 1997, 2004). Thus, this historical time frame (mid-1990s to the present) reflects the circulation of competing discourses among researchers, policymakers, administrators, school professionals, and parents regarding the best way to meet the educational needs of special education students—segregated in "specialized" settings with other students with disabilities or included among students with and without disabilities as contributing members of a general education community.

Having begun a relationship with special education professionals during Emily's preschool years in the early 1990s, Elsie recalls noticing a shift in professional language by the mid-1990s—moving away from recommendations for segregated settings and moving toward less restrictive placements.

E: So at the Board of Ed
evaluation, they said, "Well, she's seems to be on an incline." I think
that's the word that they used. "She *is* learning. We think that we
285 will put her in general ed." And at the time, I think that was when
the focus started on pulling kids out of self-contained classes—which I
was like—okay, sounds good—if you are telling me that she can make
it in general ed. I wasn't even thinking about the size of the classrooms
yet—which I wish I had. It's almost like things *had* to go the way they
290 went because of up to where we are now. "She's on an incline. She's
learning. She's never going to be at the top of her class. But she is
learning. She'll be able to make it through." I didn't look at anything. I
didn't know anything about that—like I do now. I just went with, "Oh,
300 okay." They're saying positive things—what *seemed* to be positive—
seemed. You know, they're telling you that your child is learning. She's
learning. Good! And telling you that she doesn't have to be in self-
contained class any more. Let's move her into general education. So they
asked me to sign. Everybody's saying all these positive things. She's
305 going to have resource room where they will help her with her math and
her reading. So I said, "Okay." So, I signed.

In the context of this eligibility and placement committee meeting, school professionals position themselves as experts who inform Elsie of their predictions for Emily's educational future, followed by the announcement of their decision to transfer Emily out of a self-contained setting and into a general education second grade classroom with resource services. The professionals establish authority within their use of language ("we think that we will put her in general ed" [lines 284–285]), making a declarative statement that excludes Elsie's input. Authority is further reflected in their repeated use of "we" (line 284) as well as in their phrasing "put her" (line 285)—a clear assumption of their power to determine Emily's bodily placement within the context of school.

Providing no more information than the vague notion that "she seems to be on an incline" (line 283), the professionals expect Elsie to accept their recommendation for a general education placement without question or further explanation. It is of interest that Elsie recounts the meeting without any dialogue between herself and the professionals. Instead, she relates what she was thinking in response to what the professionals *tell* her. Elsie accepts her positioning as passive recipient of expert knowledge, placing her faith in the professionals. Only in retrospect does Elsie recognize that she should have asked questions about the size of the classroom (lines 288–289), admitting "I didn't look at anything. I didn't know anything about that—like I do now" (lines 292–293). When the professionals ask for her signature, Elsie does so without question.

Elsie's exchange with these professionals effectively sustains the power relations established by a long-standing authoritative professional discourse within public schools. Elsie further muses, "It's almost like things *had* to go the way they went because of up to where we are now" (lines 289–290), alluding to the influence of a state mandate (for school districts to provide more inclusive settings for students with disabilities) that appears to influence the outcome of her individual experience.

Elsie offers additional insight into her feelings and thoughts during the first special education committee meetings that she attended.

E: I didn't have the language at the time—*not* English—the educational
or what do you call it?—the language that has to do with special ed
[**deep pause**]. Because I was so flustered at some of them and instead of—
1100 there's a saying that I repeat to Emily a lot—"Better to not say anything
and appear a fool than to open up your mouth and remove all doubt
[**light laugh**]." So, if I'm not sure, I won't say anything. People then think
that you don't know anything if you don't say anything. But I didn't want
to say something that I didn't know anything about . . . I remember the
1105 very first meeting—my *very* first IEP meeting, sitting there . . .
very quietly if you don't know much about what the process is, what
the language is. You are going to sit there very politely in the face of
authority—who you fear is authority—right? And I remember sitting
at the first meeting and I thought—here's a psychologist, here's a
1110 social worker, here's an education evaluator, here's a parent advocate—
whatever they are, I didn't know what they were at the time—and *they*
know what they're talking about. They've evaluated your kid. They've
tested your kid. And know they're going to *give* you the results. Which is
exactly the way they put it. I had met with all those people individually, so
1115 now here they are coming back with their results. I sat there very quietly,
very nicely, crossed my ankles, and folded my hands on my lap and didn't
fidget—you know, all the things that I was taught to do in an interview.
I'm sitting there very still with my heart coming out of my chest because
"What the hell am I doing here? I don't know what they are talking
1120 about!"

Elsie's narrative reveals how "the language of experts" functions to uphold institutional and bureaucratic authority (Fairclough, 1995), effectively silencing her as an outsider to "the language that has to do with special ed" (line 963). Given no entry point into the conversation, Elsie relies upon an adage, "Better to not say anything and appear a fool than to open up your mouth and remove all doubt" (lines 965–966).

Elsie's choice of language reflects her passive positioning. For example, four times within the narrative she describes herself as merely sitting there (lines 1105, 1107, 1115, 1118). She remembers feeling flustered (line 965), fearful (line 1108), and anxious (line 1118). In the "face of authority" (lines 1107–1108), Elsie responds by sitting "very quietly" (lines 1106, 1115), "very politely" (1107), and "very nicely" (line 1116). In contrast, she regards "the experts" (the psychologist, the social worker, the education evaluator, the parent advocate) as the active players in this scenario, as reflected in her use of action verbs (e.g., test, evaluate, know, give) to describe what they do. Her sense of distance from the professionals is revealed further in her reference to "they" ten times in five consecutive lines of text (lines 1111–1115).

Elsie likens her initial experiences within special education committee meetings to the experience of attending a job interview. Her analogy reflects her perception of an authoritative rather than a collaborative posture on the part of school professionals. The power imbalance in favor of professionals, who control the proceedings with their scientific language and tests, illustrates Michel Foucault's (1980) point that "knowledge and power are integrated with one another... [for it] is not possible for power to be exercised without knowledge" (p. 52). In the face of the power/knowledge nexus orchestrated by these professionals (Foucault, 1980), Elsie feels so alienated and suppressed that she wonders why her presence is required at all.

The following narrative takes place in 1997 during Emily's third grade year in a general education classroom. Elsie learns the extent of authority granted to school professionals during a particular feedback conference regarding Emily's most recent evaluation results. During the meeting, professionals speak a privileged language invested with the power to classify and define limits of possibility.

615 E: And still to this day, I do not know what
all those scores mean. I look them over and over and over again and I
wish I could just find someone to break it down to me. So that I know.
I read it and I read it and I go online. I just can't get it in my head.
Anyway, I looked at all the scores. Some were high and some were low
620 and this one was very low—72. They said it was borderline
retardation. But they just didn't understand why she was so high in these
and so low in this one—that they had never seen anything like it. They

couldn't understand. But the bottom line was that she probably would
not go to college, that she might become a receptionist, that she could
625 have one of *those* kind of jobs like that... When they said retarded, I
stopped listening at that point [**deep pause**]. I'm getting a lump in my
630 throat thinking about it. I couldn't believe that I was sitting in this dark
little room because there is no space in public school for anything. It's
always like this *closet* that we were in—this dim yellow bulb—sitting
at a round table... I'm sitting in this room which is depressing enough as it
640 is and they say borderline retarded. I wasn't prepared for that. If I'm
going to believe that, now I have to go in a *totally* different direction.
If I'm going to believe that she's borderline retarded, now what? *Now*
what? Because we're in grammar school. Is she going to get promoted?
Is she going to be able to learn? What other services am I going to have
645 to find out about? What else am I going to have to educate myself about?
Who am I going to talk to? Who am I going to have to call? Oh, my
God—what am I going to do next? It was too much. I was overwhelmed
[**voice lowers**]. I had to leave. I said, "You know. I have to go home."
That was third grade because it was Mr. Kessler. Eight years old. *Eight-*
650 *year-old.* Eight or nine because she got left back. I said, "I don't feel
well. I have to go home. I have to go home [**voice lowers**]." I picked
Emily up and we went home. I went to bed for four days. I went to bed for
four days and I didn't *think*. I was in my bed, curled up in the fetal
position, covers up to my chin. Sometimes over my head. Thank God—I
655 think my husband wasn't working at the time. He dealt with her the whole
time. I didn't want to think. I didn't want to breathe. I didn't watch TV.
I didn't want to read. I did *nothing,* but vegetate for four days. Then it was
like [**claps**]—a light bulb going off over my head. Like when you wake up
and you feel refreshed—I got up and I said, "Time for action. Now let's
660 start." Then, I started reading and I started talking. I had her evaluated
privately. And that evaluation helped me because it wasn't so *bad*, you
know. The comments and all that stuff were not as bad as what they
had said to me.

Elsie's narrative speaks to the continuing alienation of mothers whose children are described by the impenetrable "language of testing" (lines 615–617). It is of note that Elsie recalls the particular number ("72" [line 620]) that abruptly propels Emily into a life different than the one imagined moments before—a life now defined by "borderline retardation" (lines 620–621). Referring again to the professionals only as "they" (lines 620, 621, 622), Elsie recounts their puzzlement regarding Emily's extreme range of high and low scores. Despite some question about the meaning of Emily's testing profile, the professionals, exercising the ritualized nature of examination, pronounce "borderline retardation" to be scientific fact and confidently predict that eight-year-old Emily can expect to become "a receptionist, that she could have one of *those* kinds of jobs" (lines 624–625).

Like Cam in the late 1970s and Mimi, Cosette, and Marie in the 1980s, Elsie recalls a visceral response to the professionals who exercise the "great instrument of power" (Foucault, 1975, p. 184) to deem her child outside the parameters of *normal.* In the retelling of her story, Elsie resurrects deep emotion. She recalls meeting in a closet of a room with "a dim yellow bulb" (line 632)—a setting that ultimately mirrors the world closing darkly around her. While the professionals continue to talk about her child's "retardation" in their matter-of-fact manner, Elsie's mind spins with questions about how to navigate the unfamiliar future suddenly thrust upon her.

Upon careful analysis of the text, however, seeds of Elsie's eventual resistance to this powerful discourse are apparent ("*if* I'm going to believe that" [lines 640–641]), suggesting some understanding of the agency she possesses. Although she goes to bed where she remains immobilized for four days straight, Elsie emerges with a new-found conviction to challenge the discourse defining and limiting her child. From this moment forward, Elsie assumes the advocacy stance that she holds today.

In a narrative set in 1998, Chloe recounts a feedback conference in which school professionals present the results of her then 16-year-old daughter's psycho-educational evaluation. Like Elsie, Chloe experiences powerlessness in the face of an authoritative discourse that wields its scientific privilege to define, classify, and limit.

466 C: From the minute
that conference started, it was *absolutely* the most *horrendous* thing I
have *ever* gone through in my life—I'll *never* forget it! It was *demeaning.*
It was *awful.* Not just for me—because I could have taken it—but Jessie
was there. They had requested Jessie to come to it. It was just the worst.
470 First of all, the psychologist gave her report. And according to her
findings, Jessie was *retarded* [**in a flippant tone**]. She had an IQ of—
I recall one area was 74 and the other was 79. At this time again, I
wasn't *really* understanding [**speech slows for emphasis**]. Since all this
has happened, I have become the most avid reader in the United States on
475 learning disabilities and IQs and dyslexia. Anything I can get my hands
on, I read. *Now,* I'm educated on it. At this meeting, I wasn't. I didn't
know how to fight them. All I knew was what they were showing me on
paper. Jessie is not *capable* of learning. She's retarded. Her tests
showed that she is just *so* low-IQed she will never get this." This is a
kid that was making As and Bs in *advanced college prep* courses—
485 literature and sociology. She was in the top, the challenge English classes.
The *only* thing she was deficit in was math. And here she sat in this IEP
meeting being told, "You're an *idiot.* You can't learn [**draws a breath**]."

Chloe's portrayal of this conference as "*absolutely* the most *horrendous* thing I have *ever* gone through in my life—I'll *never* forget it" (lines 466–467) is a poignant illustration of a material consequence of discursive practice. She recalls "the language of experts" as "demeaning" (line 467), "awful" (line 468), and "just the worst" (line 469) not only for herself, but more significantly for her daughter Jessie who is present. Echoing Elsie's narrative, Chloe recalls the exact numbers ("I recall one area was 74 and the other was 79" [line 472]) that instantly erase Jessie's prior identity and reconstruct her as "retarded" (lines 471, 478) and "not *capable* of learning" (line 478). As Foucault (1975) reminds us,

> The examination that places individuals in a field of surveillance also situates them in a network of writing; it engages them in a whole mass of documents that capture and fix them. (p. 189)

As the school psychologist's object of surveillance, Jessie suddenly becomes captured and fixed within the official written documents around which special education spins.

In the middle of her retelling, Chloe steps out of the narrative momentarily to reflect upon the differences in her level of understanding then compared with what she knows now. With a tinge of regret in her voice, Chloe explains that she did not know "how to fight them" (line 477) and their scientific evidence of her child's supposed mental retardation. In fact, she is given no invitation to respond. Chloe's narrative is rife with language that reflects the authoritative posturing of the school professionals and the passive positioning of Jessie and herself (e.g., "the psychologist gave her report" [line 470], "according to her findings" [lines 470–471], "showing me on paper" [lines 477–478], "sat in an IEP meeting being told" [lines 486–487]).

Despite the fact that she "wasn't *really* understanding" (line 473), Chloe recognizes that such test scores make no sense in light of Jessie's record of past academic achievement and current placement and success in advanced (college preparatory) literature and sociology classes. Given that "the *only* thing she was deficit in was math" (line 486), Chloe begins to challenge the school professionals.

530 C: So, I just started talking back to the
psychologist. And the more I talked back to her, the more *nasty* she
got. Not at *me*, but at Jessie—*aiming* it at Jessie. Jessie finally
came to the point where she was bowing her head and shaking and I
knew she was sobbing. I could see the tears rolling down her face.
535 I could see that she was just in *agony*. The guidance counselor
looks at Jessie sternly and *slaps* her hand down as hard as she
could on the table like you would with a two-year-old child to get
their attention and said, "What's the *matter* with you? This is *your*

IEP meeting. *You* need to speak up and advocate for yourself! Talk!"
540 Jessie was crying so hard that she couldn't talk. There was no way that that child could utter a word. I mean, she would have been hysterical. And I looked at this woman and I said, "She *can't* talk. Can *you* not *see* that she is *sobbing*? That she's crying so hard that she can't stop?" "I'm sorry. But in high school, you have to learn to do for *yourself*. And
545 if she can't *do* for herself in high school, she's *never* going to be able to do it in life." I mean it was like a *tirade* coming down. And the more she did, the more Jessie got upset to the point that she was gasping for more air because it was all being said *about* her. She obviously could not defend herself. She *couldn't* say anything... the only thing that they
560 offered was "We can put her in with the self-contained children." I wouldn't sign that paper... I refused, but Jessie signed it. I'm sure in her records, to this day, that paper is stained with her tears because when
565 she signed it she couldn't stop crying to sign it. You know, you could barely read her signature... I truly believe that that day [**voice breaks**]
575 *something died*! I don't know if Jessie will *ever* get it back [**in tears**]. And I *knew* it that day. I could *see* it. I could see it in her *eyes* [**in tears**].

Authoritative imagery courses through Chloe's narrative. In describing herself as "talking back" (line 530) to professional language, Chloe suggests that her position is much like that of a child resisting authority—and indeed, the psychologist responds like a parent angry with an impudent child. Throughout her retelling, Chloe's language reflects the intense effort on the part of professionals to wield authority. For example, she describes the psychologist as "*aiming* [her comments] at Jessie" (line 532), who, by virtue of her student status, is the most vulnerable person at the table. Chloe recalls that the guidance counselor, assuming an intimidating body posture toward Jessie, "looks at her sternly and *slaps* her hand down as hard as she could on the table like you would with a two-year-old child to get their attention" (lines 536–538). Moreover, she uses speech to intimidate—asking Jessie a question to which no answer is expected, delivering a rapid succession of commands, and making accusatory assumptions. Chloe likens the guidance counselor's manner of speech to "a *tirade* coming down" (line 546). In contrast, Jessie's positioning reflects imagery of the powerless. For example, she is described within the narrative as "bowing her head" (line 533), "shaking" (line 533), "sobbing" (line 534), "crying so hard that she couldn't talk" (line 540), "gasping for air" (lines 547–548), unable to "defend herself" (line 559), unable to "say anything" (line 559).

It is of interest that Chloe's memory of this meeting suggests an element of confession, an integral element of the power/knowledge nexus. Foucault (1978) contends that procedures of confession operate within the regular formation of scientific discourse. Once extracted from the subject of surveillance, the confession functions as a sign that contributes to the discourse of truth. Indeed, the scenario Chloe paints bears resemblance to a police interrogation in which a suspect is pressured to confess. Jessie, the focus of investigation, becomes the target of intimidating and rapid-fire questioning from the authorities. Under the pressure of such interrogation, Jessie becomes uncontrollably upset and unable to defend herself. Jessie, falsely accused and intimidated, signs the special education documents put before her—a kind of forced confession of mental retardation—"the complement to the written, secret preliminary investigation" (Foucault, 1975, p. 38). In exchange for this admission of "guilt," the authorities offer segregation from society ("we can put her in with the self-contained children" [lines 560–561]), a "bargain" that Chloe refuses.

So great is the material consequence of this particular discourse upon Jessie's life that Chloe believes that "that day [**voice breaks**] *something died*! I don't know if Jessie will *ever* get it back [**in tears**]. And I *knew* it that day. I could *see* it. I could see it in her *eyes* [**in tears**]" (lines 574–575). In a later part of the interview, Chloe explains that another psychologist subsequently verifies Jessie's cognition as within the average- to high-average range and identifies a learning disability in mathematics. Chloe notes that this knowledge does little to restore Jessie's self-esteem. She laments that Jessie, now a young adult, lacks focus and direction in her life.

The mothers of this generational group report intense discomfort regarding the objective, decontextualized, and depersonalized stance of professionals. For example, Katie recalls a special education committee meeting in which professionals do not know the gender of her child.

K: All they're looking at is the *name*—[the child's real name] is an unusual
310 name. They didn't even realize he was a *boy*! They just *assumed* that
[child's real name] was a girl's name. *I* had to interject—excuse me—it's
a *boy*. "Oh! I'm sorry [**in a sarcastic tone**]." Well, you'd think they'd
know this going in. Don't you think they'd look a *little* closer at the forms
in front of them?

In that this exchange takes place at the onset of the meeting, Katie immediately distrusts the professionals' degree of sincerity and investment in her child. She mocks their apology and expresses disbelief at the blatant inattention to her child's identity.

Elsie recounts a special education committee meeting in which she makes a conscious move to resist professional objectification of her child. As the meeting opens, Elsie disrupts the usual discursive practice by passing Emily's photograph to each person at the table.

E: I learned
510 that they don't know who the student is, so I brought a picture. They

passed around the picture. "Oh yeah. Beautiful child." Blah, blah, blah. The usual condescending stuff that they say. Patronizing, not condescending. "Oh, that's nice." Why would I want to look at a picture of your kid? You know? So they could see who Emily was. That she was not somebody in their minds—I don't know *what* was in their minds—*I* wanted *them* to see that she was well-groomed; *I* wanted *them* to see that she was good-looking; *I* wanted *them* to see that she
520 has a beautiful smile—that she was a *real* kid. I can't believe the amount of paperwork that they have at these meetings... So they're looking at all these papers and they're sitting there [**demonstrates rustling papers**] shuffling papers, shuffling papers, you know, smiling at me the way that
525 they do.

Elsie's increasing sense of agency is evident within her narrative. For example, she repeats the phrase "*I* wanted *them* to see" (lines 518–519) three consecutive times, reflecting a marked shift in her positioning. In her more proactive stance, Elsie challenges whatever ideas the professionals might have about her daughter by introducing Emily's photograph. Resisting conceptualization of her child as nothing more than a compilation of test scores, Elsie demands a contextualized perspective of Emily as "a *real* kid" (line 520).

It is of particular interest that Elsie portrays the professionals in passive terms, a significant departure from her earlier descriptions. In contrast to her prior use of action verbs to describe what professionals do, Elsie now depicts them as "looking at" (line 522), "sitting there" (line 523), and "shuffling papers" (line 524). No longer intimidated by the authoritarian discourse, Elsie mocks what she perceives as the professionals' condescending and patronizing attitude ("blah, blah, blah" [line 511]).

In a narrative set in the late 1990s, Kim describes her ongoing frustrations with the "language of experts."

K: It seemed like
390 what they wanted to talk about mostly were all the *test scores* and I really didn't *care* too much about them because what I knew one thing—he didn't test well [**light laughter**], so I didn't care. I didn't put a lot of stock in that. I knew another thing—he *was* a good student and he *was* going to make it, so what *I* was most interested in was what we can do
395 to make sure he *not only* stays afloat, but gets a good grasp of the material while feeling good about himself while learning enough to pass on to the next grade. I didn't really want to go over all those tests. And I remember it taking up *so much* time. And I remember looking at my watch thinking—"Oh, gosh, they only gave me 30 minutes and I've
400 not said half of what *I* wanted to say"... And *all* they wanted to do was talk about the *percentile* and the *battery* and you know, the numbers, and I didn't really care about the numbers and I still don't.

Kim's narrative illustrates how the privileging of professional language functions as a barrier to collaboration. She describes the professionals as imposing their agenda upon the meeting by speaking only about "test scores" (line 390), "*percentile*" (line 401), "the *battery*" (line 401), and "the numbers" (line 401), thereby silencing her contributions and point of view ("I've not said half of what *I* wanted to say" [lines 399–400]).

Kim resists "the language of experts" in her dismissal of its relevance to her child's educational growth and self-esteem (lines 394–395), claiming authority in her own knowledge of her child (lines 391–393). She perceives the professional agenda as "taking up so *much* time" (line 398), leaving little room for her own perspective and concerns (line 402).

In the following narrative, Elsie explains how she learns to engage proactively with "the language of experts."

1120 E: So as I learned, I got out of that role at each subsequent meeting—
as I learned more and I realized—especially when I decided to *really*
take matters into my own hands. Finally at this meeting that I'm
talking about now—I had to *visually* put myself in that place where
I am *now* going to come in and you're *not* going to just be condescending,
1125 you're *not* just going to *tell* me what it is that *I* need. *I* know what we
need now. *I* know that I can say what I want and *know* that I can be
vocal. I know that they can't just—I know *some* of the language—so
I know *now* that they just can't say "Well, we're going to do this and
this and this and two times a week and two times a week and this
1130 many kids in the class—blah, blah, blah [**spoken at a frenetic pace**]."
Now I know when they say 10:1 what that means, 3:1 what that means,
3:1:2 what that means, you know. You had the kids in speech and
language—you had the teacher and the group which is 3 and the other
time it was 1:1 or whatever. Blah, blah, blah. 1:1, 2:1, 3:1, blah, blah, blah
1135 *whatever* [**spoken at a frenetic pace**]. So now I know. So now I *know* all
this stuff and I *thought* I knew *so* much. Now at *this* point, I know that
I only knew *this* much, you know [**laughter**]? But I feel like I know *this*
much. I'm coming in and out, I'm ready, you know! Yeah—I'm
Sigourney Weaver! I'm coming in!! Rock and roll! Here I come
1140 [**laughter**]!

A notable shift in agency is evident within Elsie's narrative. For example, she uses the pronoun "I" 31 times in 21 lines of text. Action verbs appear in her language, such as "learned" (lines 1120, 1121), "realized" (line 1121), "decided" (line 1121). Moreover, the phrase "I know" appears 11 times within 12 lines of text (lines 1125–1137).

Elsie is able to move out of a passive role after gaining access to "the language of experts" (line 1127). Conscious of the connection between knowledge and power (Foucault, 1980), Elsie confidently asserts, "You're *not* just going to *tell* me what it is that *I* need. *I* know what we need now" (lines 1125–1126). She mimics the professional language that once intimidated and alienated her, frenetically rattling off numbers and words common to special education jargon (lines 1128–1135). In her repeated use of "blah, blah, blah" (line 1134) and "whatever" (lines 1134–1135), Elsie reveals her derision toward professional language. Moreover, she compares her newfound aggressive persona to Sigourney Weaver's bold and rugged female character in the science fiction film *Alien*.

Conflicts in Shared Decision Making and IEP Implementation

IDEA guarantees parents the right to collaborate with professionals in making educational decisions regarding their children with disabilities. However, mothers of this generation document persisting tensions between parents and professionals in regard to shared decision making.

Reflecting back upon more than a decade of engagement with special education professionals, Elsie reports less than satisfying collaborative experiences.

E: I refer to it as
The Inquisition. They are all sitting there. They all look like stone statues.
Actually, it's kinda like [**pause**] scary and funny at the same time
because they're sitting there with these smiles on their faces and their
680 heads are nodding, you know? Whenever you make a comment or ask
a question, they'd say, "Um-hum [**demonstrates nodding head and
smiling insincerely**]." It's like when you go to an old age home and
you're sitting there with a friend or relative or somebody you may not
know because I volunteered at old age homes. You're going, "Oooh,
685 how *beautiful* you look!" And you know what their reaction is going to
be because they're sitting there—nobody's visiting them—and you go
in and you say, "How *beautiful* you look, Miss So and So!" And they're
beaming! So, they're going "hmmmm" to me at this meeting—like I'm
going to *beam* at the comment that they're going to give me. I'm *not*
690 beaming [**laughter**]!! I want to get up and chop your heads off! One at
a time! Lop them off! Just sitting there in order—just go chop, chop,
chop, chop—and watch them roll down on the floor [**demonstrates
and laughs heartily**]!

In comparing special education committee meetings to The Inquisition, Elsie suggests that her experience of shared decision making is analogous to facing a tribunal dedicated to the suppression of heresy. It is worth noting that Elsie's metaphor mirrors Foucault's (1980) assertion that "power never ceases its interrogation, its inquisition, its registration of truth" (p. 93). Her characterization of school personnel as "stone statues" (line 677) reflects the alienating effect of their objective and impersonal professional stance. Moreover, Elsie observes that professionals respond to her in much the same way that visitors patronize nursing home residents. She wryly comments that professionals expect her, like a senile patient, "to *beam* at the comment that they are going to give me" (lines 688–690). Indicative of her growing resistance to authoritarian discourse, Elsie shares a playful fantasy in which she usurps professional power. She goes on to speak about a real-life special education meeting in which she disrupts and redistributes power.

695 E: So I came in. I was prepared *and* I was very nervous because I'm thinking, "Now what's going to happen?" I knew that they couldn't do anything to me, but I was still nervous. So I come in and I sit down. I have it in my lap, waiting for the opportunity to say that I'm going to bring out the tape recorder. "I'm so and so. I'm the psychologist. I'm so and so. I'm
700 the social worker. I'm so and so. I'm the ed evaluator. I'm so and so. I'm the parent advocate." I go, "I'm so and so and I'm taping this meeting [**laughter**]." I can still see that. It's like slow motion. Like when somebody falls and it's in slow motion. Everybody looks at each other. Then they look at me. Still the frozen smiles on their faces. I started
705 taping. But then there was *no talking*. You hear all the papers shuffling. Papers are being shuffled. Very minimal comments are being made. I don't remember what the end result was at that meeting. But [**slight pause**]—it was a useless meeting.

In anticipation of challenging school authority, Elsie remembers feeling very nervous. By introducing a tape recorder into the meeting, Elsie shifts power away from professionals and toward herself—as only an informed parent knows of the right to tape record and how to use tape recordings for accountability purposes. Elsie recalls how the professionals wordlessly exchange glances with one another, then resist her abrupt redistribution of power by simply not talking.

Reflecting upon her evolving relationship with special education professionals, Debra describes the process by which she shifts from passive participant to passionate advocate.

D: Because the funny thing is, as an adult, you remember when you were a child and having to be obedient to your teachers. And even as an adult,
205 you sometimes you feel like you have to sit there and listen to the teacher! Because the teacher tells you something and it's "yes, ma'am" even though I'm 41 years old now. You have to get over that and find it in yourself to say—you know what? No. No. Wait. I'm educated. I'm an adult. I don't have to sit there and say "yes ma'am" any more to the
210 teachers. Not that I don't respect them and respect what they're trying to do. But again, it's realizing *you're* the only voice for your child. And you have to be heard. I think when you start to see that your child is coming

> home very upset *every* day, that starts to get you angry. You start out being upset. You start out crying with your child when they cry over their
> 215 homework. You cry *with* them. Then that starts to turn to anger and frustration. You realize as his parent—*I'm* his voice—he doesn't have a voice. He's a child in school and they have to say "yes, ma'am" and do whatever they're told. Well, *I* don't have to do whatever I'm told. *I'm* his mother. You get to that point of being so *angry*. You know there's the
> 220 straw that breaks the camel's back. For me, it was that fourth grade teacher telling Justin that she was disappointed in him. That's when we did a total flip and I was *so* angry. That was *it*. That was the final straw.

Having attended public schools herself, Debra attributes her initial passive participation to an ingrained obedience to teacher authority. To "get over that" (line 207), Debra engages in self-talk (much like Cam and Rosemary describe in chapters three and four) to claim and assert her own authority ("I'm educated" [line 208]; "I'm an adult" [lines 208–209]; "I don't have to sit there and say 'yes ma'am' any more to teachers" [line 209]; "well, *I* don't have to do whatever I'm told" [line 218]).

In mapping her progression toward advocacy, Debra recalls grief over her son's pain ("you start out crying with your child when they cry over their homework" [lines 214–215]), followed by "anger and frustration" (lines 215–216), which eventually turns into action ("you realize as his parent—*I'm* his voice—he doesn't have a voice" [lines 216–217]). It is noteworthy that Debra (like many of the mothers in chapters three and four) identifies a turning point ("the straw that breaks the camel's back" [lines 219–220]) that represents the specific moment in which she shifts to a proactive advocacy stance.

In response to authoritative rather than collaborative posturing on the part of school professionals, Kanene challenges the taken-for-granted procedures that characterize decision making in special education committee meetings.

> K: And the psychologist says, "Oh, this is what your child needs." And I'm saying to myself, "Now, I'm at home with him. And the teacher's in the classroom with him.
> 145 Me and her is the only two really dealing with him." And when they try to tell me, "Oh, your son needs this." I'm saying like, "He don't need this." "Well, he need another evaluation." "I'm not having him used as a pincushion." That's basically what they were wanting. "Well, maybe we can get this done." "No more. No. My son is a human being
> 150 just like you. The only thing I'm asking is to get him the services. That's all." You know, believe me, they stare at the psychologist. And all of them are sitting there talking to me and like I told them, "I don't even know who *you* are! Because you're not sitting here with your MD showing me anything. You're just sitting here with your idea. Anybody can go and get

155 an idea." You hit a nerve with them as well as they hit a nerve with you. But you gotta let them know you gotta prove to me that you're this type of person. "Where you work at? What are you doing? Are you *really* a psychologist? I don't see no papers in front of me. I don't know *who* you are!" What was they telling me? Something about brain waves they
160 wanted me to have. I said, "No. No, no." So they'll send you to all these testing. No. "You put *your* child through that and tell me how *you* like it." But you gotta let them know, "I'm not *stupid*!" They tell me, "Well, *we* think." "No, you can't *feel* like I feel because he's with *me*, not with *you*. You're coming in here to tell me what you got on that paper. Are you
165 in school with him?" It feels like I'm being *dominated* as a parent, you know, like in other words, "Oh well, we're going to do this regardless of what. We are going to do it *our* way." You have to let *them* know.

From her perspective as a parent, Kanene questions the logic that positions a school psychologist, the person with the *least* contextual knowledge of her child, in the role of "expert." As she explains, "I'm at home with him. And the teacher's in the classroom with him" (lines 143–145). Kanene dismisses the psychologist's "scientific" assertions about what her child needs, thereby expressing contempt for what professionals value as knowledge.

In response to the team's recommendation for additional evaluation ("something about brain waves they wanted me to have" [line 159]), Kanene resists "the oppressive practices of the objectification of human beings" (Foucault, 1980, p. 238) by refusing to have her son "used as a pincushion" (lines 147). Asserting that her son "is a human being just like you" (lines 149–150), Kanene challenges the professionals to reflect upon the consequences of their practices by putting themselves in her position ("you put *your* child through that and tell me how *you* like it" [lines 161]). In much the same way that Cam, Rosemary, and Debra reclaim their knowledge, Kanene refuses to accept a passive role in decision making, asserting her capability for contribution: "I'm not *stupid*!" (line 162).

Kanene disrupts the professional expectation that parents accept their authority without documentation or explanation of what constitutes their expert status, turning the tables by asking questions of the professionals: "Where you work at? What are you doing? Are you *really* a psychologist? I don't see no papers in front of me. I don't know *who* you are!" (lines 157–159). Furthermore, Kanene defies the taken-for-granted assumption that professionals, privileged in their socially sanctioned position of authority, know what is best for other people's children to the exclusion of other ways of knowing: "No, you can't *feel* like I feel because he's with *me*, not with *you*. You're coming in here to tell me what you got on that paper. Are you in the school with him?" (lines 163–164). She sums up her experience within special education committee meetings as "being *dominated* as a parent" (line 165).

Echoing concerns of the two previous generations of mothers, Debra points out that educational decisions agreed upon within special education committee meetings are not always implemented consistently at the classroom level.

D: We started this last year and we've decided to do it every single year. We call a meeting about two weeks after school starts. At least, the teachers have met Justin. We want *every* teacher there that he has on his team because in seventh grade he has six or seven teachers. We want them
135 to know all of the disappointments we've had in the past. The way teachers have treated Justin in the past is why we have decided it's no longer going to be the school taking charge. The parent has to take charge. We have to just lay out the plan for the teachers in the beginning. We have to call numerous meetings. Maybe three or four meetings throughout the
140 school year so that *they* know we're still paying attention.

Given "all of the disappointments" (line 135) she has experienced over the years, Debra distrusts school personnel to monitor her son's education. Her decision "to take charge" (line 137) reflects the absence of any meaningful collaboration with the school. For example, Debra takes lone initiative "to just lay out the plan for the teachers" (line 138), following up with "numerous meetings" (line 139) throughout the school year.

Elsie likewise reports disappointment regarding classroom implementation of educational decisions. The following narrative takes place after Emily moves from a self-contained setting to a second grade inclusion class.

E: I was there *every* day and *asking* him every day, "How is she doing?" "Oh she's doing okay. She shared so and so today. She's doing fine." I remember once I said—I looked at some of her work—"I think she needs
320 a little help." "Well, you know, I have the *other* kids." Blah, blah, blah, So. I said okay. When I went for report cards—"Oh, she's doing fine." They were just patting me on the head. "We're not expecting much" is what I see now. That's what they were telling me. "Let's not expect much. She's doing okay." So at the end of the year I look at her books
325 and her notebooks and her workbooks—and there's *nothing* in there! I'm thinking to myself, "Oh my God! She's been in school for a whole year, hardly any homework—which some teachers don't give homework so I really didn't think anything of it. It's second grade, you know. A little sheet—write the letters whatever. But she hasn't done *any*thing.
330 She's been sitting in the classroom doing *what*? She's done nothing! *Nothing*! Because here is the evidence of it! There's *nothing* in her books!! Maybe something on *one* line! You know? The daily journals that they had to write. She had her name and not even correctly. And I was like—we are *not* going to go through another year like this. *No*.

It is worth noting that Elsie's 1990s recollection bears striking resemblance to Mimi's 1970s account of attempts at collaboration with her son's kindergarten teacher (chapter three). Having had her concerns repeatedly dismissed, Elsie, like Mimi, becomes convinced that the teacher is providing an appropriate education for her child. From the vantage point of the present, Elsie realizes that assurances about Emily's adequate performance were, in fact, relative to the low expectations held for her: "Let's not expect much. She's doing okay" (lines 324–325). Within this "inclusive" placement, Emily's educational needs become construed as less important than the needs of "the *other* kids" (line 320) who presumably belong. After discovering that Emily has "done nothing" (line 330) the entire school year, Elsie vows to take sole responsibility for monitoring her child's education.

In the following narrative, Kanene expresses misgivings about her decision to place her son in a special education class.

K: I remember when Marvin was in elementary
school. I walked into his special ed classroom to see what he's doing
because he wasn't coming home with no homework. He wasn't coming
home with anything. I'm looking outside the classroom door and there
475 he's sitting in the corner, pulling thread off the carpet. And the teacher's
teaching the rest of the class. I was not happy with that. That gave me
grounds to go back to that office and tell them, "Get my son out." How
you gonna take, "Well, he was just sitting there!" "You're a special
education teacher. Why you just cannot say, 'Marvin, you got to
480 come over here and join us.' He's a seven-year-old child. What you
mean he don't want to participate? Get over there! Are you telling me that
these kids are doing what they want to do?" You got to stay on top of
things with them. It's a *hard* thing. I'm telling you, it's *hard*. It's not easy.
It doesn't *get* easy. It hurts when you're dealing with it. It *hurts*.

Anticipating that her child's educational needs will be met within a special education placement, Kanene is stunned by the low expectations she observes in this classroom: "I'm looking outside the classroom door and he's sitting in the corner, pulling thread off the carpet" (lines 474–475). Kanene comes to understands that her son is no longer seen as the seven-year-old child she knows, but rather as a "special education student" whose disability affords him less accountability than other seven-year-olds. Like Elsie and Debra, Kanene concludes that she must "stay on top of things" (line 482) in order to ensure an appropriate education for her son. She laments the difficulty, pain, and constancy of this task: "It's a *hard* thing. I'm telling you, it's *hard*. It's not easy. It doesn't *get* easy. It hurts when you're dealing with it. It *hurts*" (lines 483–484).

In the following narrative, Kim explains how teachers respond to implementing accommodations on Jacob's IEP.

K: At one point,
165 I had four kids in the same elementary school. So, I had an advantage in that I had a good relationship with the school prior to the point that I really needed them to work with me. And I had a good rapport and they knew me as someone that was helpful and not just someone there to complain. I felt like that was a plus for me and something I had on my side. And
170 even with that, I started to feel *pushed back* a little bit. I started to feel like they didn't want to see me coming like they used to *really* want to see me coming because they knew I was going to help *them*. But all of a sudden, I was coming *not* to help put up their bulletin boards but to say, "You know, Jacob needs his test orally. Or I need copies of his notes. We didn't
175 get complete copies. Or we have a test and he's not really clued into what to study. Or he *knew* this material! We studied for weeks and did all these flashcards and he knew it verbally and he made a 30 on the test. We've got to help pick him up. He's down right now."

It is noteworthy that Kim perceives implementation of IEP accommodations as a continual process of negotiation rather than a guarantee under the law. As a longtime school volunteer, Kim views her "good relationship with the school" (line 166) and "good rapport" (line 167) as "a plus for me and something I had on my side" (line 169), a kind of a bargaining tool that would enhance the probability of teachers *choosing* to implement the accommodations. Despite drawing upon her class privilege as a stay-at-home mother available to volunteer (Brantlinger, 2003), Kim soon senses teacher resistance to her shift from teacher's helper to parent advocate: "I started to felt like they didn't want to see me coming like they used to *really* want to see me coming because they knew I was going to help *them*" (lines 170–173).

Katie likewise recounts an example of teacher resistance to implementing the accommodations on her son's IEP.

K: There was one particular
science teacher that pretty much blew me off. I didn't realize she blew me off. But, the way she blew me off was—at the parent-teacher
825 conference at the open house, I went and visited every single one of his teachers and introduced myself and told them that I hoped they were aware that Louis had an IEP with certain accommodations that needed to be met—and one, in particular, was that he needed to be seated at the front of the class. She proceeded to tell me, "Oh, it really doesn't matter
830 because I rotate throughout the whole class. And no matter where he sits, at some particular point, he won't be where I am." And I thought, "Well, that's okay. If she does wander, that's helpful to all kids, so I'm not going to complain about that." Well, *then* it came to my attention through Louis,

that she *does* stay at the front of the class during lectures. The only time
835 she wanders is when they are doing experiments. So, it *did* make a difference where he was seated during lectures. So, I proceeded to write a nice note. I did. I thought I tactfully said that due to the conversation that Louis and I had had that you do wander during experiments, but at lecture time you do predominantly stay at the front of the class and it would be
840 most helpful if you would move Louis to the front. She moved him. And I didn't have any more problems. But, I thought—the way she just handled the whole situation, I felt like she was just trying to blow me off. Like—Okay—here's a parent that mothers her child too much [**pause**]. "And, you know, I'm not going to be catering to obsessive mothers [**laughter**]!"

It is of interest that the science teacher appears to regard her responsibility for IEP accommodations as a matter of teacher discretion, rather than a guarantee under the law. Katie recalls the teacher being dismissive in response to seating Louis at the front of the classroom. In an effort to persuade the teacher to move Louis, Katie writes "a nice note" (lines 836–837) in which she "tactfully" (line 837) suggests that "it would be most helpful" (lines 839–840) for Louis to sit in the front of the classroom during lectures. Despite her son's right to accommodations under the law, Katie feels she must *appease* the teacher in order to gain access. Moreover, she suspects that the teacher, regarding her as "a parent that mothers her child too much" (line 843), has no intention of "catering to obsessive mothers" (line 844)—supporting my earlier assertions that school professionals may perceive mothers of children labeled LD as overinvolved and enmeshed with their children.

Elsie reports a similar teacher response to her proactive involvement in Emily's education.

805 E: Then she went to junior high school which was a *nightmare*. I did my thing that I usually do. I went the first day of school. This time, I wrote letters. I wrote a letter: "Dear Teachers, I'm so and so. My daughter is so and so. I'm very involved. Anything—here's my number. Here's my email. Here's this. Here's my address. You can reach
810 me. You can write me. Here's a book. Please write your comments. I'll write my comments. Any questions—let me know." Every which way you could do it. And I *know* that I was perceived as possessive again.

In light of pervasive laments of school professionals regarding the lack of parental involvement particularly within urban schools, it is worth noting that Elsie becomes pathologized as possessive because she *is* involved in her daughter's education. Echoing Elsie's comments, Kim explains how mothers become positioned in this no-win situation.

K: It's that Mother Hen Syndrome. When you're trying to do it for them and trying to help them so much because *you have to*—because you're

820 never *sure* the IEP is going to be followed—then you feel this *urge* that you've got to cover your tracks, their tracks and his tracks, too, because you *don't want him to fail.* Then people are like, "Oh, you're overinvolved. You're overprotective."

Given reluctance on the part of school personnel to assume responsibility for her child's education, Kim becomes a "Mother Hen" (line 819)—not because she *wants* to but because she feels she *has* to. Thus, Kim suggests that mothers become pathologized as "overinvolved" (line 823) and "overprotective" (line 823) by the very people responsible for creating their anxiety ("you're never *sure* that the IEP is going to be followed" [lines 819–820]). In light of her cogent analysis, it is of particular interest, then, that Kim accepts rather than resists the pathologizing of her son as a classroom burden.

180 K: And *even* now when I have teachers with my children that came along after Jacob and it's like—I know they're good teachers, so I'll say, "Oh, I hope Rebekah gets you next year!" And I even feel the need to say, "All the work you put in with Jacob will be repaid because she's so independent and she's such a strong student, you won't
185 have to do any extra with her." I feel the need to *say* that so that they're not going, "Oh, *not* her again [**light laughter**]!" "We don't want *her* as a mom!"

Despite her son's right to a free and appropriate public education, Kim perceives classrooms as *really* belonging to strong, independent students such as her daughter, Rebekah. By telling teachers that "all the work you put in with Jacob will be repaid" (line 183), Kim implies acknowledgment of her son as a burden requiring extraordinary effort on the part of teachers—a *favor* for which she owes them. In fact, Kim appears to offer her daughter, Rebekah, a student who does not need teachers "do any extra with her" (line 185), as *compensation* for the "extra work" created by her son's educational needs. It is of particular interest that Kim feels "the need to *say* that" (line 185), as if to assure the teachers that she *can* conform and be a "normal" mom of a "normal" child.

Katie, on the other hand, considers how teacher attitude contributes to the construction of her son as "a problem."

K: I think the conscientious teachers don't view him any different than
815 any other student. I think the *lazy* teachers *do* view him differently and not that he's not as smart, but *they* have to put forth a better effort on their presentation of the material. They have to put forth a little more effort on the teaching procedures. They have to put forth a little bit of effort in making sure that they present the material

820 in the manner that *he's* going to grasp and the other ones, too. So, the lazy teachers view him as "*a problem.*"

Katie understands that a difference exists between "conscientious teachers" (line 814) and "lazy teachers" (line 815) not only in their perception of Louis, but also in their response toward him. For example, Katie observes that within the context of a conscientious teacher's classroom, Louis is viewed no differently than any other student. In contrast, Louis is considered to be "a problem" (line 820) within the context of a lazy teacher's classroom. Katie suggests that Louis becomes construed as a problem when teachers do not put forth effort in their instruction, a point she emphasizes three times (lines 816, 817–818, 819). Thus, Katie's observations serve as an example of disability as a social construction (Corker & Shakespeare, 2002; Linton, 1998; Ware, 2004).

In a narrative that takes place in the fall of 2003, Debra explains how competing discourses—specifically the 1997 reauthorization of IDEA and *No Child Left Behind* (2001)—construct her son in different and mostly incompatible ways.

D: At an IEP meeting with all
110 of his teachers in the very beginning of the year, we shared our concerns, past bad experiences with the teachers, and things we didn't want to happen this year. A *week* later, I get a letter from his homeroom teacher saying that we have to come in and sign a letter for a Plan of Action because Justin may not pass this year. So I called her on the phone and I
115 said, "We just had an hour long meeting last week with all of his teachers for a new IEP and planning everything. *Why* am I being called back in a week later to sign *this*?" "I didn't know you had a meeting last week. No one asked me to be there. We could have done all this at the same time." So I went in to meet with her and I said, "You do know Justin has a
120 learning disability and an IEP." "No, I had no idea." So once *again*, we're presented with the checklist of "If Justin doesn't do all his homework and doesn't read 20 minutes a day and this and that, then he may be left behind this year." Even *though* he is *labeled* LD with an IEP [**deep pause**]. So, I don't understand what the school system is doing. I
125 don't get it. You get lost at that point [**voice lowers**].

Debra's narrative illustrates not only a lack of intersection between federally mandated discursive practices, but also among school professionals who implement mandates at the local level. Operating under two sets of federal guidelines, school professionals simultaneously lay claim to Justin's education without apparent knowledge of one another's agendas. While IDEA guidelines guarantee an IEP on the basis of his specific educational needs as a student labeled LD, *No Child Left Behind* (2001) holds Justin accountable to educational standards expected of *all* students. Although *No Child Left Behind* (2001) purports to build upon the 1997 reauthorization of IDEA, these local school professionals remove themselves from accountability by requiring Debra to sign a prepared statement at the *beginning* of the school year that makes Justin responsible for his failure to learn: "So once *again*, we're presented with the checklist of 'If Justin doesn't do all his homework and doesn't read 20 minutes a day and this and that, then he may be left behind this year'" (lines 120–123).

Debra's account reflects my earlier argument that the standards discourse engenders a "personal responsibility" rhetoric reminiscent of welfare reform, a viewpoint that extends the blame for school failure onto students and their families. In other words, if students "choose" not to rise to the standards expected, they will suffer the consequences—and in this case, ironically, the consequence is being "left behind" (line 123). Debra questions the logic that holds Justin accountable to standards without consideration of his *school*-documented learning disability: "[**Deep pause**] so, I don't understand what the school system is doing. I don't get it. You get lost at that point [**voice lowers**]" (lines 124–125).

Devalued Knowledge of Mothers

Narratives from this generation, like those of the previous two generations in chapters three and four, feature professionals who routinely privilege expert knowledge. In particular, mothers continue to report instances in which their early concerns are dismissed by professionals. Elsie illustrates this point below.

E: I knew she should be making certain sounds at a certain age and it wasn't happening. She had a speech and language eval. The results were—"She's
180 too young. Let's wait a year. Bring her back in a year. It's just that she's so little. Can't really tell [**slightly sarcastic tone**]." Blah, blah, blah, blah blah. I'd like to go back and strangle all of them [**laughter**]! Because it was a wasted year when—*in my gut*—I *knew* that there was a problem.

In this scenario, professionals give precedence to scientific knowledge (test results) over Elsie's observations and intuition about her own child. In the absence of "scientific data," professionals are unable to act upon Elsie's concerns and recommend follow-up testing in a year. From the perspective of a mother who sees her concerns played out in her child's *daily life*, Elsie questions the logic of withholding intervention until Emily's challenges can be verified by a *test*. In Elsie's mind, this amounts to "a wasted year" (line 183) in the life of her child. From the vantage point of the present,

Elsie mocks professional language (lines 179–181) and muses, "I'd like to go back and strangle all of them [**laughter**]!" (line 182).

As a regular parent volunteer in her son's first grade classroom during the late 1990s, Kim recalls her early concerns about Jacob's slow acquisition of reading skills. In the following recollection, Kim describes his teacher's response to her.

K: I started spending a lot of time in the classroom and one of my jobs in the classroom was to have kids read to me. That's really when I began to notice that *a lot* of these kids were reading by mid-first grade. He was nowhere *near* reading. That was my first clue, I guess. The
80 teacher was very much—"Don't worry. He's doing great. He'll read. Late bloomer. Late maturity. He's going to read. He's going to be fine." The teacher really *adored* Jacob and had such an affection for him. She saw how *enthused* he was with learning and she saw those same stars in his eyes that I had seen. He was a child the teacher loved because he wanted
85 so bad to learn—kinda like—well, so what if he's not reading—he *will* because he is such a good student. He *didn't*.

This scenario illustrates how a teacher privileges her professional experiences with struggling readers over a mother's observations and instincts. In her confident prediction of Jacob's future success with maturity, the teacher fails to consider any other possibilities that might account for Jacob's academic struggles, thereby dismissing Kim's concern that her son was "nowhere *near* reading" (line 79). Given Kim's position as a classroom volunteer, it is also possible that the teacher simply views her as an overanxious mother comparing her child's performance to that of other children.

In a narrative remarkably similar to Kim's, Katie recalls expressing concerns to her son's kindergarten teacher.

K: Even as early as preschool, I was seeing some problems with Louis. Then, he went into formal education at a private Catholic school which was kindergarten where they were teaching him his letters and the sounds of
15 the vowels and consonants. He could grab that. He could tell you what the *letter* was, he could tell you the *sound*, but he couldn't put them together into any way, shape, or form of making words. So, we just thought that was just a developmental thing. I was a *little bit* concerned over that, but the teacher kept telling me not to worry—"He's a *boy*. Boys are just a
20 *little bit* slower and *not* to worry." There *was* something needling at me in the background, but I was trying *not* to be an *overly* concerned mother of a first child. I just took her at her word and kept going on and on.

Although Katie articulates a cogent analysis of Louis's reading difficulty, the teacher discounts this information in favor of her professional experience that suggests all boys are "just a *little bit* slower" (line 20). Aware of her vulnerability to being perceived as "an *overly* concerned mother of a first child" (lines 21–22), Katie takes the teacher "at her word" (line 22), subsequently dismissing her *own* knowledge despite "something needling at me in the background" (lines 20–21). Upon entering grade school, Louis continues to struggle with reading and writing.

K: So, he goes into first grade. He's making A/B honor roll, but yet *I* can *see* a problem with the fact that he can't seem to read. When I'd
25 actually have him sit down and read, it was a chore. He was taking *forever* to do his homework. But yet when I would quiz him myself on things, he could spout everything off. He was just *incredible.* But then, if I made him actually *write* it or look at a sentence—which in first grade, you're supposed to be *leaning* toward the direction of reading—he was not
30 even getting *close.* I was concerned that the teachers weren't concerned. They were telling me, "No, he's doing great. According to his grades, he's performing, he's performing, he's performing." But by the end of the year, when they're making these children rely more on their *own* ability to read, *then* they started seeing roadblocks. So, here I've got a second
35 grader who can't read *a lick* and before even the *first* nine weeks are up, the teacher is already talking to me about having him held back.

Katie recalls her eroding confidence in expert knowledge during Louis's first grade year. She questions the discrepancy between the school's evaluation of Louis's performance ("he's making A/B honor roll" [line 23]) and her own observations of Louis ("*I* can *see* a problem with the fact that he can't seem to read" [line 24]). Katie grounds her well-articulated concerns in the kind of everyday, intimate knowledge mothers possess about their children (Belenky, Clinchy, Goldberger, & Tarule, 1997). Yet, the teachers resist and dismiss Katie's input in favor of their expert perspective. Katie recalls how "they were telling me" (line 31)—reflecting her positioning by teachers as a receiver rather than contributor of knowledge. Katie's repetition of the teacher response ("he's performing" [line 32]) three consecutive times suggests an air of parental authority—akin to the familiar "because I said so" response meant to silence a questioning child.

Katie recalls that the teachers finally recognize that Louis "can't read *a lick*" (lines 34–35) at the end of the school year when first graders are expected to engage in independent reading. As a material consequence of their practice, Katie faces a second grade teacher who considers Louis to be a probable candidate for grade retention "before even the *first* nine weeks are up" (line 35).

Echoing the preceding narratives of Elsie, Kim, and Katie, Debra remembers expressing early concerns to teachers regarding her then three-year-old son's speech development.

> D: I would always ask the preschool teachers, "Do you notice anything different in Justin from the other kids?" And it was always, "Well, boys develop slower. Don't worry about it. He's fine." And I would say—now, of course, he's almost
> 20 *three*—"But, he can't say his name. He can't say Justin clearly." "No, boys are slow. Don't worry about it. Everything will be fine in a couple of years." His doctors were never concerned about it because Justin had asthma and that always took precedence over anything else. The doctors were always more concerned about the asthma than any concerns we had
> 25 about his speech.

It is noteworthy that Debra's retelling includes multiple instances of absolute and binaric language. For example, she uses the word "always" four times (lines 16, 18, 23, 24) and the word "never" once (line 22). Debra's word choice to describe her experiences with professionals suggests authoritarian rather than collaborative exchanges.

Debra recalls being placated by professionals ("don't worry" [lines 18, 21], "he's fine" [line 19], "no, boys are slow" [lines 20–21]). The preschool teachers confidently predict that "everything will be fine in a couple of years" (lines 21–22), thereby dismissing any other possibilities that might account for Justin's speech difficulties. Debra fares no better with medical professionals. She remembers doctors being primarily concerned with her son's asthma to the exclusion of "any concerns we had about his speech" (lines 24–25).

In a latter part of the interview, Debra explains how she acts upon her own instincts and seeks a private evaluation. Results of this multidisciplinary evaluation indicate dyspraxia and a speech/language disorder, making Justin eligible for speech therapy services under IDEA guidelines. Although he consistently engages in early intervention services throughout preschool, Justin experiences difficulty in kindergarten.

> D: When Justin started kindergarten, he was always behind *everybody*. The teachers would *still* say, "Boys develop slower. Don't worry about it. He'll be fine in a few years." I distinctly remember *really* noticing him falling
> 40 behind in first grade. First grade, he could *not* keep up *at all* with *any* of the work.

Despite a dual diagnosis of dyspraxia and a speech/language disorder, Justin continues to be dismissed by his teachers as simply "immature" (lines 38–39). Debra, on the other hand, persists in expressing her concern that Justin is "always behind *everybody*" (line 37) and unable to "keep up *at all* with *any* of the work" (lines 40–41). It is worth noting that "immaturity" requires *no action* be taken on the part of teachers other than waiting for the child to age. Later in the interview, Debra credits Justin's second grade teacher for recognizing the probable relationship between a speech/language disorder and academic difficulty. Results of a subsequent evaluation make Justin eligible for learning disability resource services.

Influence of Race/Culture, Social Class, and Gender

Unlike the singular perspective of White mothers in chapters three and four, this generation of mothers represents greater diversity. As stated earlier, Kim, Katie, and Debra are White, although Kim and Debra define their cultural backgrounds as Appalachian and Italian American, respectively. Elsie is Hispanic and of Puerto Rican heritage. Kanene is African American.

Recalling her early experiences in special education committee meetings, Elsie considers the impact that racial identity may have had upon her initial interactions with professionals.

E: I'm Hispanic.
960 There was no one Hispanic in there [**pause**]. That's not true. The psychologist was Hispanic. And *maybe* the parent advocate at one of them. But even so—I think it did come into play [**deep pause**]. Because I was so flustered and if I'm not sure, I won't say anything. People then think you don't *know* anything if you don't *say* anything. But I didn't
965 want to say something that I didn't know anything about. So, they *assume* things about you.

Despite the presence of one or possibly two Hispanic members of the special education committee, Elsie believes that her race "did come into play" (line 962). Initially flustered by unfamiliar special education language and procedures, Elsie listens rather than speaks. She intimates that professionals may have interpreted her silence as incapability—in part because of her Hispanic identity.

In contrast, Kanene does not believe that her racial identity has any bearing upon her relationship with professionals.

K: They do it to *all* races. No. No. No. If they can hold on to that little bit of money, you could be *purple*. If that district can, they
390 will take that money and utilize it for a computer.

On the basis of her lived experience, Kanene perceives special education as a commodity that school personnel withhold from parents and begrudgingly parcel out only when forced to do so. From this standpoint, Kanene sees race as having nothing to do with a parent's success in accessing services ("you could be *purple*" [line 389]). In fact, she emphatically states that "they do it to *all* races" (line 388).

On the other hand, Elsie, living in the same metropolitan area as Kanene, has noticed that parents of color increasingly resist rather than pursue special education services.

E: The first reaction is only
1390 what parents have heard in the media—that only Blacks and Hispanics are in self-contained classes and they are warehousing them. That seems to be—and it *was*—that *is* what was happening. So they are afraid. The minute they say self-contained class—whether it's good for them or not—they're like, "No way! I'm not doing it!"

Elsie's comments reflect the circulation of a discourse gaining momentum—specifically, that special education is separate and unequal, particularly for the disproportionate percentage of African American and Hispanic children who populate its ranks (Artiles, 1998; Hartocollis, 1998; Losen & Orfield, 2002). Elsie points to "the media" (line 1390) as the primary conduit through which parents access this discourse. As parents share with one another reports of "warehousing" (line 1391) Hispanic and African American children in segregated special education classrooms, Elsie sees increasing refusal of special education services among fearful parents who see special education as an educational dead end for their children.

From her standpoint as a White mother of adopted children with diverse racial and cultural backgrounds, Kim shares her perceptions about the interplay between race and special education.

K: Well, first of all, I'm White and middle class. So, I probably [**slight**
515 **hesitation**] was just like everybody else—you know—the school personnel in the meetings—and didn't really see it as a big issue. Never really thought about it until—well, several different thoughts here—until I learned *differently* [**light laugh**]. The first being when Jacob first went to his resource class, he came home and said, "Mom!"
520 and he starts naming *all* the little boys in there with him. And I remember them well from kindergarten because they were *always* the troublemakers, always the ones that never had any lunch money—or I guess they were on free lunch—and always the ones that couldn't come to the class party because the mama wouldn't come to drive them and they were all the
525 *dark* boys. And I looked at it and I said, "You know—how come Challenge is full of all the White kids—and I am *not* prejudiced, but that's just the way it was—and all the dark kids go to resource?" And Jacob noticed that. Not that it bothered him. We have different races within our family of adopted children. His *brother* is dark! But, it was
530 definitely something he noticed.

As previously discussed in chapters three and four, public discussion of race, for the most part, is considered unseemly within traditional White

Southern culture. Kim's discomfort becomes evident within the first five lines of text. For example, she reveals her tentativeness in the use of the word "probably" (line 514), after which she hesitates (lines 514–515) as if weighing the appropriateness of her words before speaking them aloud. She obliquely alludes to her White privilege, relying upon the euphemism of being "just like everybody else" (line 515)—meaning other White people such as "the school personnel in the meetings" (lines 515–516). From a White perspective, Kim notes that "it" (lines 516, 517)—meaning racial inequity—was not a "big issue" (line 516) or even something she ever "really thought about" (line 517). It is of particular interest that Kim implicates school personnel, whom she identifies with as White and middle class, as being equally unaware of racial inequities, a point that supports Brantlinger's (2003) assertion that White, middle-class teachers are "unaware of their complicity in systems that stratify and oppress" (p. 5).

Once her son enters special education, Kim loses her privilege to remain unaware. Upon realizing that Jacob's special education classmates are "all the *dark* boys" (lines 524–525) who are "always" (lines 521, 522, 523) poor, excluded, and considered troublemakers, Kim suddenly recognizes that Challenge (the gifted program) is "full of all the White kids" (line 526). As if to justify making this statement aloud, Kim adds, "And I am *not* prejudiced, but that's just the way it was" (lines 526–527). Her observations support the ample documentation of underrepresentation of children of color in high-status programs such as gifted and talented programs and their overrepresentation in low-status classes such as special education (Artiles & Trent, 1994; Barton & Oliver, 1997; Connor & Boskin, 2001; Harry, 1992, 1994).

In reflecting upon her continuing experiences within special education committee meetings, Elsie suggests that a parent's ability to negotiate for quality special education services may be impacted by professional assumptions about his or her social class as well as cultural background.

E: I think it was my cultural
background because of what I've seen from my *own* observations.
A lot of the people who go in who have kids in special ed self-
contained classes are Hispanic and Black. I saw very little White kids.
980 *Very* little. And why is that? Because they have the resources, okay?
Most of the time, I suppose. I'm generalizing and that sounds biased
also—which is really bad. But if you're sitting there with a lawyer,
then you have resources. The rest of us—we're not coming in with
lawyers. I think that you can make the assumption that then either we
985 don't know or we don't have the money. So that's not far fetched. Okay? I
didn't come in dressed up. *Now* I know that I don't come in jeans and
sneakers. Okay? Because that first impression is everything. And that's the
only impression they ever have. They never saw *me*. And if they saw me

at another meeting—which I did meet *some* of the people at other
990 meetings—they didn't *remember* me. You know? Those factors do
come in.

Informed by her "real world" observations, Elsie contends that Hispanic and African American children populate self-contained special education classes, in part, because their parents do not have the same economic resources as White parents who can afford to hire lawyers to negotiate on their behalf. Elsie implies that White parents with resources command more attention from school professionals than do Hispanic and African American parents (Harry, 1992, 1994) without comparable resources. Moreover, Elsie believes that professionals may make initial assumptions about parents based upon their manner of dress. Reflecting upon the multiple layers of professional assumptions with which a parent must contend, Elsie laments, "They never saw me" (line 988).

Drawing upon her experiences within the same school district as Elsie, Kanene believes that a parent's social class, regardless of race or cultural background, has the greatest influence upon negotiations with school professionals.

K: I've seen White parents go through it just like me. Unless
you got money. Then you can get a lawyer. If you got money, you got
that type of lawyer. Now if you're rich, regardless if you are a rich Black
395 person or a rich White person, you got money and you got a lawyer, yeah,
they gonna move faster. People who don't have much got to stay on top of
these people. You got to keep fighting.

Kanene contends that parents of higher socioeconomic status not only have the resources to secure legal assistance, but also are more likely to have an already established relationship with a lawyer. She agrees with Elsie that parents of higher social class, given their resources and legal connections, command attention from school professionals; on the other hand, parents of lower socioeconomic status must *work* at being noticed and heard in order to garner attention ("people who don't have much got to stay on top of these people. You got to keep fighting" [lines 395–396]), thereby confirming Delpit's (1995) contention that people without class privilege are acutely aware of the benefits afforded those who do possess class privilege.

While reflecting upon the evaluation process for special education, Elsie considers the scope of personal information required as part of the institutional procedure.

E: I just thought of something. They do a whole social eval, too. They do
all this social background so they *know* Emily was adopted, what my

1160 educational level is. They *know* what my economic level is. They *have* all that information. So you can *make* assumptions based on those things. If somebody comes in and they're on public assistance, it's there. If you have ten kids, it's there. If you live in a one room house with 50 million people, it's there. You know? Well, I never thought about it before. I
1165 know enough to know that a lot of those things play into [**slight pause**] the makeup of yourself or your child. It does. It *has* an impact. It has an impact if you are in a one-room apartment and you have five kids living there. It has an impact if you are on public assistance and the way that you live your life. It has an impact. But what they can't do is make
1170 *assumptions* based on that information—which they *do*—you *know* they do.

In the midst of the interview, it occurs to Elsie that school professionals may make assumptions not only because of her cultural background, but also because of the demographic data collected and placed within her child's special education file. Professionals determine *what* data to gather from parents and for *what purpose*—an example of a "type of power which is constantly exercised by means of surveillance" (Foucault, 1980, p. 104). Given the inherent power differential between the interviewer (the school professional who controls information) and the interviewee (the parent who gives information)—conceptualized by Foucault (1973) as "those who unmask and those before whom one unmasks" (p. 110)—Elsie considers how this particular kind of data collection, extracted out of the *context* of individual lives, might contribute to professional assumptions about parents. Thus, Elsie recognizes that the "turning of real lives into writing…functions as a procedure of objectification and subjection" (Foucault, 1975, p. 192) not only for her child, but also for herself.

Not unlike the two previous generations, mothers of higher social class report continued reliance upon private educational resources. Despite the promise of a free and appropriate public education for all children, White middle-class mothers of this generational group access private resources to supplement their children's education. For example, Kim and her husband send their son to a private school for students with learning disabilities because of their growing dissatisfaction with special education services.

K: I joked and said that I would have cleaned toilets in the middle of the night if I had to send him there. It is *that* good and that strong. Of course, now I've spent all my money on private
280 school [**light laughter**]—I don't know where his college money is going to come from! But at that point in sixth grade, Dave and I looked at each other and said, "We don't have it. We don't have the money for a private school." And Dave said, "Kim, we spend his college money right now or else he's not going to *need* it."

Despite never having envisioned private school for any of their five children, these middle-class parents are able, nonetheless, to afford private school tuition by using money they have set aside in a college fund. Kim, a stay-at-home mother, jokes about having said she "would have cleaned toilets in the middle of the night" to afford the tuition (lines 277–278)—an exaggeration meant to reflect her level of determination. However, it is worth noting that it *is* an exaggeration precisely because most middle-class women would never *have* to consider such a job. Although certainly a financial stress, tuition is not prohibitive to this middle-class family; therefore, private school education is an option available to them—yet another example of the intersection between disability and social class.

Katie, who works part-time from home to be available to her four boys, comments on her oldest son's private tutoring.

K: And *luckily*, we were able to afford it. It was a *burden* on us, but we
110 agreed that it was worth the money we were having to pay for him to
get the private tutoring in order to help him to get up to the standard
of where he was supposed to be in the classroom. Because, I mean,
he wasn't even *close*. You know, he's just starting to meet up with
yesterday's standards and they're continuing on. So, we felt it was
115 worth it.

In the way that she frames having access to private tutoring ("and *luckily*, we were able to afford it" [line 109]), Katie implies some awareness of class privilege in her appreciation of their monetary resources to do so—but leaves unsaid that other parents might not be as lucky. Much like Della in the 1960s, Katie relies upon a private tutor to support her child in a classroom context that fails to address his educational needs. It is of interest that Katie refers to the expense of private tutoring as a "*burden*" (line 109) that is "worth the money" (line 110) without challenging why her child's educational needs cannot be met within the classroom.

In contrast, Elsie, a working mother, lacks the financial means to secure private resources as a supplement to Emily's public education.

450 E: So I was out working. And then I was working with
Emily. So, it was Emily and me. Emily and me. Emily and me. Emily
and me *all* the time. Which I really think led to our divorce, you know
[**voice lowers**]. Without even thinking about it in *those* terms, that was
where we started to go apart, then, because it was just like, I have *no time*
for anything else but to pay attention to this.

In the absence of outside resources, Elsie assumes the role of full-time tutor and devotes herself to Emily's education, a commitment that leaves "*no time* for anything else" (lines 454–455) including her marriage. In fact, she considers her eventual divorce to be a material consequence of her all-consuming focus upon Emily's education.

Echoing narratives told by mothers in the previous two generations, mothers of this generation observe the impact of gender upon their interactions with school professionals. In the following narrative, for example, Kim describes how the presence of her husband influences the dynamics of an IEP meeting.

K: I didn't notice it at all until Dave just happened to go with me to an IEP meeting and I could see *right away* that it was a *totally* different setting. It went from being—you know—little neighborhood mom who's coming in for PTA in
540 her jumper doing the fund-raising activities—to being a *business* meeting. The language was different. It was more professional. We stayed on track. We didn't chit-chat. We didn't have to talk about everybody—you know about what everybody did on the weekend. We went in, we had an agenda, we got down to business. He had his business suit on and they
545 knew *he* had *important* places to go, so the meeting was totally different.

In response to a male presence, school professionals (identified earlier in the interview as all women) conduct a business-like meeting—in striking contrast to the casual and conversational atmosphere of prior meetings. Observing this "totally different" (line 545) meeting, Kim becomes acutely aware of a significant discrepancy in the degree of respect commanded by her husband's gender in comparison with her own. In somewhat sardonic terms, Kim suggests that school professionals assume her husband, a man in a business suit, has "*important* places to go" (line 545)—implying that the kinds of places where "men" work are more important than the elementary schools where "women" work—while she, on the other hand, is positioned by school professionals as "the little neighborhood mom" (line 539), obedient, available, and eager to carry out school activities prescribed for Good Mothers—a gender role assignment seemingly far out of sync with the present, but nonetheless consistent with the history of parent and professional relationships within public schooling outlined in chapter two.

Reflecting upon her ongoing interactions with school professionals, Kim recalls being routinely dismissed in regard to her concerns and/or requests.

560 K: I was treated *totally* different. I learned *right* away that if Dave went in his *business suit* on his way to work that he could go in with *one* request, *one* reason, and he would state his request and he would leave. And he would go, "Kim, I don't know why it's such a big deal! I'm in and

out in ten seconds. Why won't they listen to *you* [**laughter**]?" So, then I
565 began feeling like I had an inferior complex like "You're right, Dave. This *is* easy. This is *not* complicated. You state your request and you leave." But, I never felt like I could do it [**voice lowers**]. So, I started depending on Dave—like he needed *one* more thing to do and again, this gets back to—this was *my* job. Dave's job is to go earn the money. My job is to take
570 care of the kids. And here I couldn't even do it because they wouldn't *listen* to me. So, I would have to interfere with *his* time and send *him* in, get *him* out of work to come to the meetings because when I came with Dave, I was *always* treated more professionally than when I came alone [**voice lowers**].

Recognizing the advantages inherent to male privilege, Kim relies upon her husband, Dave, to parlay concerns and requests to school personnel. Never having experienced the world without the benefit of male privilege, Dave appears baffled that Kim is unable to command results as he does—affirming research that suggests men are largely unaware of the advantages inherent within their masculine embodiment (Connell, 1998; Gerschick, 1998; Haraway, 1988; Harding, 1998; Lorde, 1984). In turn, Kim becomes convinced that her ineffectiveness results from personal shortcomings, rather than the ways school professionals *position* her to be ineffective. As a material consequence of such positioning, Kim develops feelings of inadequacy about herself as a mother ("Dave's job is to go earn the money. My job is to take of the kids. And here I couldn't even do it because they wouldn't *listen* to me" [lines 569–571]). At two different points in the interview, Kim lowers her voice as if ashamed to acknowledge her perceived inadequacies aloud.

In a narrative that takes place in 2002, Katie recalls observing the influence of gender roles during her first IEP meeting, scheduled a few weeks before the due date of her fourth child. When the baby arrives earlier than anticipated, Katie, determined to attend the meeting, leaves the hospital with her newborn and goes directly to the school.

K: I literally walked out of the hospital and went to
435 an IEP meeting [**laughter**]. And they were just all thrilled when I walked in with this newborn. Awwww—and they were making over him. That was sweet. Everybody was really nice, but, I guess because I'm a mother with a newborn infant, they figured, well, she can't possibly contribute to this meeting—which *he* was perfect—he sat there and didn't
440 utter a peep the whole time. So after they oohed and aahed over the baby, they all immediately turned and focused on my husband. It was very *funny* in that they automatically looked at *him*. They automatically directed *everything* at him. They are doing *all* of the interchange across from *him*. Even though I was there, I didn't feel like they even *cared* about
445 my opinion. Nothing was ever asked of *me*.

In response to a mother holding a newborn, school professionals construct a particular identity for Katie—that of a Madonna-like figure consumed by infant devotion—her embodiment serving as a repository for cultural assumptions about selfless motherhood (Grosz, 1993; Warner, 2005; Young, 1990). Having been assigned this singular identity, Katie becomes positioned as unable "to contribute to this meeting" (line 439)—a position she notably accepts rather than resists. It is of particular interest that no comparable identity is projected onto the father of the newborn. Katie observes how professionals automatically direct their eye contact, attention, and conversation toward her husband, entirely ignoring her presence.

K: But [**pause**] I felt more
455 like an observer to the meeting than a participant in the meeting. And
[**slight pause**] I guess because so much was going on, I was just *so grateful* that this meeting was taking place that I didn't even sit back
and—I never harbored any resentments or was shocked at—I really
didn't even think about it—to be honest—[**slight pause**]—until later.
460 Here I had gone through all kinds of hoops with this child, worried
and fretted, and pretty much approached any problem that he had *with* him side-by-side, and [**pause**] I guess this is kind of egotistical, but
there's Ray getting *all* the credit for it [**voice lowers**].

By opening her statement with "but" (line 454), Katie signals that what follows will reveal what she really thinks. After pausing for a moment, Katie states that she felt "more like an observer to the meeting than a participant" (lines 454–455). It is of interest that Katie pauses again, then offers several somewhat disjointed statements that partially explain and/or justify her passivity during the meeting—revealing a certain tentativeness about expressing what she *really* wants to say. Following another pause, Katie finally claims her anger about the blatant lack of regard for her knowledge, experience, and commitment. Yet, she retracts from her rightful indignation by suggesting in a lowered voice that she might be "kind of egotistical" (line 461) for resenting "Ray getting *all* the credit" (line 463)—an example of women's cultural conditioning, particularly in traditional White Southern culture, to accept rather than resist their second-class status to men. It is worth noting that Ray readily embraces the male privilege school professionals afford him, thereby contributing to rather than challenging Katie's passive positioning.

Conclusion

During this historical time frame (mid-1990s to 2004), the institution of special education is well established within the public school system. Yet, a growing number of critics question the efficacy and ethics of what has become a separate and largely unequal system of education. In response to evidence of special education students routinely being excluded from the general education curriculum, the law's 1997 reauthorization requires that special education students have accessibility to the general education curriculum. Thus, competing discourses circulate among researchers, policymakers, administrators, school professionals, and parents regarding the best way to meet the educational needs of special education students—segregated in "specialized" settings with other students with disabilities or included in general education among students with and without disabilities.

Despite the presence of a growing ideology about inclusive practices, mothers continue to report ways in which their children routinely become excluded within public education. Moreover, these mothers, like those of the previous generations, attest to school professionals who dismiss their potential to collaborate meaningfully on the basis of their motherhood. Drawing upon their experiences within the institution of special education, mothers describe being passively positioned by an authoritative discourse that claims the power to define, classify, and limit their children.

More than 30 years after the passage of PL 94–142, tensions persist between these mothers and professionals. Each mother in this generational group describes both engaging with and resisting against authoritative discourse; however, mothers who represent nondominant culture (e.g., African American, Hispanic, at or below poverty level) consider dominant culture mothers (e.g., White and middle to upper class) as better able to command attention from school professionals because of their inherent social, economic, and cultural capital. Yet despite the advantages that dominant culture mothers possess, *all* of the mothers relate experiences that lead them to distrust school professionals; subsequently, mothers assume primary (rather than shared) responsibility for their children's education.

Chapter Six

Mothers Speaking to Mothers: A Cross-Generational Conversation

As described in chapter one, the five mothers who participated in the group interview represent a subset of the 15 mothers interviewed individually. All reside within a particular geographic region of a southeastern state. I recruited members of this group on the basis of their availability and willingness to participate. I had intended to recruit a diverse group with respect to race/culture and class; however, as I describe more extensively in chapter one, this goal proved more difficult to achieve than originally anticipated. Thus, all members of the group are White and represent the middle- to upper-class socioeconomic range. However, the group composition does reflect each generation of mothers as defined in the study. Mimi and Lindsey, whose respective children were born in 1968 and 1977, represent the first generation; Sue, whose child was born in 1984, represents the middle generation; and Kim and Katie, whose respective children were born in 1989 and 1990, represent the most current generation (see appendix F).

With the exception of Kim and Katie who consider themselves acquaintances, none of the mothers had met prior to the group interview. Yet, the mothers readily engaged in conversation with one another—finding points of connection, listening compassionately, and validating one another's perceptions and feelings. Having already interviewed each mother at length, I took note of which narratives each mother chose to share with the group. I observed that mothers told these narratives to the group with the same intensity of emotion and the same language—nearly word-for-word in most cases—as had been told to me.

As a way to initiate and circulate conversation among the participants, I began the group interview by asking each mother to draw a sketch representing her past and/or current experience within the special education system. In turn, each mother shared her drawing with the group and explained its meaning. This group conversation is represented in the transcript below and notated as described in chapter one (see appendices F–J to view a copy of each mother's drawing).

[Katie shows her drawing to the group]
Katie: This is supposed to represent how I feel about my child and
what we've been through. Here's what I drew. I drew a racetrack.
At the end of it is the finish line and Your Child's Life
Achievements. You got The Normal Child's lane. Both start
5 off at equal places in life, I feel like, okay? Here are the starting
blocks. And here's The LD Child's lane. You see that this is
supposed to represent hurdles that they have to keep jumping
over
Kim: That's so cool
Katie: to try and keep up with their peers. And
10 here's the cheering section. This is The Normal Child and this
is The LD Child. There's *all* kinds of people cheering that Normal
Child on. Helping them, supporting them, guiding them.
Kim: And cheering that *mom* on, too!
[group laughter]
15 Katie: Yeah, cheering the mom,
on, too, for doing such a good job! And there's The LD Child.
And he just has piddly two people over there. That's how I feel. I
feel like there's *so* much for The Normal Child and there's just so
few people out there that grasp what The LD Child needs.
20 Lindsey: And they
don't even *see* all these hurdles.
Katie: They don't! And they don't really
care!!
25 Lindsey: Uh-uh!!
Katie: That's the *hard* part. The people that don't
care **[voice lowers].**
[group murmurs in agreement]
[Kim shows her drawing to the group]
Kim: And your comment **[to Katie]** is *exactly* how I picture it.
30 I couldn't draw anything because the emotion that I felt from the
the moment that I first realized something was different here to
this day is just *tears*. I would go to the meeting and I would have
my agenda, have support—I would have the goals—I would do
everything I was supposed to do. And I would walk away with my
35 head down and *bawl*. *Every single time!* And it's so true! It is so
hard to stand up there and have them take you seriously when you
are that emotional about it. Of *course*, you are emotional about it!
It's your *child*!! And it's being afraid that they're not going to
jump that hurdle or have that cheering section.
[group murmurs]
40 **[Lindsey shows her drawing to the group]**
Lindsey: You know, you go into school with this little flower that you grew
and you hold it up and then, there's the principal—you know—he

represents education. I don't know that he was *the* principal, but he represents education. You know, they all have this funny hair
[group laughter]

45 Lindsey: and he has this big smile that you know is really a *wicked* smile. And these are the teachers lined up here. These are the bright children over here. The Sunshine Children.

Katie: Ooooh.

Lindsey: And this one right here is looking at her watch and saying, "I
50 really don't have *time* for this." And she's already going in another direction because the "A" Children are up here. These are going, "I don't know what you're talking about!" Here are those big red X's they put all over his papers. And another teacher is going, "What?!?" And these are the Ds and Fs. This is the
55 conglomeration of bringing to school a *perfect* person and then *this* is your reception.
[mothers murmur among themselves]
[Sue shows her drawing to the group]

Sue: This is *my* guy. This is Evan. When I embarked on this long journey with him, I had just been divorced so it was just the two of us. You have a child—he's a *beautiful* kid, he's really
60 funny, he's real bright. But he won't read. Can't read. Bangs his head on the counter. Grades are horrible. He's acting out. He's doing weird things in the classroom. I felt very isolated. I always felt like "Where do I turn?" And the reason I drew it this way is because I felt like it was just Evan and I, hanging on
65 together, trying to hang onto whatever air we could to figure out the pressure, the turbulence around us. We were always going to do it together. I always felt like we were just *underwater*. There was nowhere to turn. It was dark. Just isolated. When you speak to people, they can't hear you. They *don't* hear you. So,
70 that's why I drew it this way. Because it was just the two of us, hanging on together, hoping that we're going to make it.
[group murmurs in response]
[Mimi shows her drawing to the group]

Mimi: Well to understand my picture you have to understand that we came along—Pete was born in 1968—*before* Public Law 94–142. So we were actually—probably I shouldn't say it in this way—but
75 I felt like a trailblazer. What *we* did constantly was go to meetings. It was one meeting after the other. So when I think about the *horrors*—well, there's many things—but one of the things that came to mind for some reason was sitting around a table at a school district office with *all* of these people trying to make a
80 decision about what to do with Pete. We would go to staffings and they wouldn't just bring the teacher and the area special education coordinator. They would bring everybody. So we would go to meetings with the teacher and the district special ed coordinator and the area special ed coordinator and the nurse and
85 the psychologist and the principal and the area superintendent and

the social worker and Harry and me.
[as Mimi points out all the people at the table in her drawing, the group audibly gasps]

Mimi: And that's all of us sitting around the table. And then here's me with tears rolling down my face remembering all that we had been through with our child since he was born.
[group murmurs in support]

Mimi: I told them all about my child. And it was like I had cast my pearls
90 before swine. Literally cast my pearls before swine. *They did not care.* What I *needed* so much was for someone to *listen* to me.

An analysis of the transcript reveals four themes that characterize this cross-generational conversation about mothers' experiences within the special education system: (1) exclusionary practices within schools; (2) feelings of isolation; (3) alienating professional practices; and (4) professional indifference. In the discussion that follows, I address each theme within the context of the mothers' drawings.

Exclusionary Practices within Schools

Katie draws a racetrack representing the school experiences of two types of children—"The Normal Child" and "The LD [learning disabled] Child"—who compete against one another to reach the finish line of "Your Child's Life Achievements." Although both children "start off at equal places in life" (lines 4–5), Katie depicts school as a race stacked in favor of The Normal Child who runs without the hurdles that obstruct The LD Child's lane. Katie's racetrack is a striking metaphor for the parallel and unequal systems of general and special education (Brantlinger, 1997; Losen & Orfield, 2000). The distinction between The Normal Child and The LD Child is exemplified further by two separate cheering sections. Katie contrasts the status of The Normal Child, a highly valued student who belongs and has "all kinds of people cheering" (line 11), with the low status of The LD Child symbolized by the "piddly two people" (line 17) in his cheering section.

Lindsey likewise depicts school as a place for some children but not for others.Her drawing reflects the open hostility that school professionals direct toward the "little flower" (line 41) that symbolizes her child. She explains that the principal, a looming, menacing figure with a "wicked smile" (line 46), represents how she views public education (lines 43–44). Teachers are depicted as frowning, checking their watches ("I really don't have *time* for this" [lines 49–50]), and clueless ("I don't know what you are talking about!" [line 52], "What?!?" [line 54]). Big red X's and Ds and Fs rain down upon her child while teacher attention is directed toward

The Sunshine Children (line 47) floating safely above the fray and among their As.

Feelings of Isolation

Sue's underwater scene, in which she and her son grope their solitary way through dark waters, symbolizes their journey together through an educational system that offers no guidance, support, or personal interest. She explains that her drawing represents "just Evan and I, hanging on together, trying to hang onto whatever air we could to figure out the pressure, the turbulence around us" (lines 64–66).

Katie's drawing likewise represents her feelings of isolation. She contrasts a crowded and enthusiastic cheering section for The Normal Child ("there's *all* kinds of people cheering that Normal Child on. Helping them, supporting them, guiding them" [lines 11–12]) with a nearly empty cheering section for The LD Child.

Alienating Professional Practices

In a simple, stark line-drawing, Kim depicts her experience of special education committee meetings. Four faceless, floating heads represent the objective and distant professionals. In contrast, Kim draws herself, a more fully human stick figure with hair, arms, and legs, walking away from the table with tears streaming down her face. In a context where rationality prevails (Shildrick, 1997), Kim feels shame about having emotion ("and I would walk away with my head down and *bawl*" [lines 34–35]) as well as resentment about being denied expression of her emotions ("of *course*, you are emotional about it! It's your child!" [lines 37–38]).

Mimi's drawing also depicts a special education committee meeting. It is noteworthy that Mimi labels each of the ten professionals present at a meeting that took place nearly 30 years ago. Full facial features appear only on her husband and herself. Other than a social worker who bears just a grimace and the superintendent who has a mouth and devil horns atop his head, Mimi, like Kim, draws the objective professionals without facial features. While the professionals shuffle their papers in the middle of the table, Mimi, tears rolling down her face, reflects upon her child's life represented in the balloon above her head. The sun, symbolizing the bright humanity of her child, contrasts with the professional focus upon the scattered papers.

Professional Indifference

Referring to the disinterested professional response to the story of her child's life, Mimi describes how she "cast her pearls before swine" (line 89). She explains, "*They did not care.* What I needed so much was for someone to *listen* to me" (lines 90–91). Sue likewise comments upon her difficulty in gaining the attention of professionals, comparing the experience to being unable to speak or hear underwater: "When you speak to people, they can't hear you. They *don't* hear you" (lines 68–69). Echoing Sue and Mimi, Katie laments about the "people that don't *care*" (lines 26–27) whom she represents as fans supporting The Normal Child.

This group conversation suggests that mothers, speaking from within their respective generations, share a striking commonality of experience within the special education system. Moreover, the mothers, while listening intently, audibly and nonverbally (e.g., head nods, laughter) confirm the resonance of one another's narratives within their own lives.

Following the drawing exercise, the group hungrily shared experiences with one another. Given my background in facilitating parent groups, I anticipated that the mothers would regard the group interview as a rare opportunity to tell their stories to a compassionate and interested audience. A good portion of the transcript includes a repetition of narratives told by mothers during the individual interviews, thereby confirming the importance of these particular stories within their lives.

Later conversation segued naturally into a discussion about how the school system might better meet the needs of parents whose children receive special education services.

Mimi: And in listening to you tell your story
[to Lindsey] about graduation and seeing that we *all* responded the
215 way we responded to it and we were *all* heartbroken at the same
time—what there is such a need for is for a *support group.* But
it can't be a whining, complaining support group. I've seen those
in my years. I've seen those support groups. Everybody goes and
complains about the school district. Somebody emerges as the one
220 who is going to be their advocate to fight the school district and the
whole thing becomes so adversarial. Everybody becomes so angry
that all of their good energies are turned inward toward this anger
that they have. But there's such a need for people to be able to sit
around and just have somebody *listen* to their story. And bottom
225 line is that there is a need for a support group or a support system
even in which there is a coordinator

Lindsey: Maybe I've just found my new
colleague!
[group laughter]

Mimi: A coordinator that sets up groups in different communities.

230 Sue: Because it's almost like—does a tree make a noise if it falls in
the woods and there's no one there to hear it?

Mimi: That's exactly right.

Sue: Because if you are fighting by yourself, there's no evidence.

Mimi: Just think about if you were in a group—and it needs to be all
235 ages, people with children who are grown and those who are
just coming along—because just think if you had been in a
group and you had someone who had already been through
what you went through with the graduation, you could have
called that person and they could have heard you out and
240 hopefully they would have been wise enough to tell you some
wise things. I sort of know what the school system can do and
what its limitations are. I mean I really do. They have serious
problems with funding and serious problems with being
understaffed and all that. And you can't *make* teachers do what
245 they don't want to do. But if people had their *own* support
systems to help them, I think that would make all the difference
in the world.

Lindsey: I had that same feeling, Mimi, when I was thinking about what
would I suggest. Before you go to an IEP meeting to have a
250 mom—not an *angry* mom—but a mom who says, "Now this is
what's going to happen. And this is what this means. And let's
think about some of the questions you want to think about. Let's
talk to the psychologist ahead of time. Let's look at the testing.
Let's see what your child's needs are. So when we go in that
255 meeting, you can contribute. Because who knows the child better
than you do?" Instead of going in there with twelve people and
you, all by yourself, the school language is not going to make
any sense unless you happen to be an educator or you happen
to have read a book. And you do whatever they say—but to have
260 someone with a desire to know the whole other side of our
story. But, I think it would be like creating Mothers Against Drunk
Drivers, the MADD thing, if we did it the wrong way. It would be
anger, holding onto the grief, it would be resentment, and it would
be all negative. If we created a group that was going to be where
265 we held each other up and we helped each other by being
constructive, that could be wonderful.

Mimi: And as a school board member, interestingly enough, the word
began to spread around—"What is *she* doing on the school
board?" Well, *she* has a handicapped child. And so what began
270 to happen was that I would get calls from these people who had
problems. Then I had fulfilled what I had hoped I could fulfill
when I ran for school board—which was to help other people
and to help the school district do better toward parents.

Lindsey: You know, *all* of us have told about how in some way we helped
275 somebody else. And I remember in second grade, after they
had started a reading book over for the third time and it's February,
and we're on page one for the third time, same reading series—
and I finally said to the principal, "Okay. Now you listen to me,
but what about the *other* kids in that group? Have they been tested?
280 Is anybody over here talking to you about *them*? Why are *they* not
reading?"

Kim: You know what I think would be an awesome idea? If you could get maybe say all the eighth graders that are getting ready to transition and have it like a master class—you know, a long class,
285 as a class for kids that would be resource kids to teach them self-advocacy. How to have the confidence and understand your disability enough to ask for what you need. You know, wouldn't it be great if the kids could be *over here* being taught by someone, you know, that knows how to teach those skills and *over here*—
290 if the moms wanted to come together and find out ways that we can support each other. Just help each other through those transitions.

Reflecting upon the compassion and validation shown for one another in this group, Mimi suggests the need for parent support groups within local communities. Given her standpoint as a former school board member, it is worth noting that Mimi's vision specifically does not include whining, complaining, adversarial, and/or angry parents, but rather parents of all ages who wish to sit and listen to one another's stories as well as participate in information sharing and mutual support. (It is of interest to consider that school professionals labeled *Mimi*, at one time or another, as a whining, complaining, and adversarial parent.) Sue affirms Mimi's suggestion in a way that reflects her previously articulated feelings of isolation ("because it's almost like—does a tree make a noise if it falls in the woods and there's no one there to hear it?" [lines 230–231], "because if you are fighting by yourself, there's no evidence" [line 233]). Citing problems with district funding and staffing as well as the inability to "*make* teachers do what they don't want to do" (lines 245–246), Mimi focuses attention upon ways of helping parents to cope *within* the present system, rather than advocating for systemic school change. Moreover, she conceptualizes parents creating "their *own* support systems" (lines 245–246) without input from or collaboration with the professional community, a point of particular interest to me in light of the success I experienced with parents and teachers creating mutual support networks within a shared group setting.

Lindsey concurs with Mimi and extends parent support to include mother-to-mother advocates who act as guides through the special education system. Lindsey, like Mimi, envisions a supportive and constructive network that discourages negativity, anger, grief, and/or resentment—leaving uncontested the system that engenders such feelings among parents. Thus, Lindsey likewise fails to challenge the *need* for parent advocates within a system intended to embrace parent participation.

As the conversation ensues, a slight shift occurs as the group suggests that responsibility lies not just with parents, but also with school personnel. For example, Mimi acknowledges how she used her school board position (i.e., power) to "help other people" (line 272) negotiate the special education system as well as to hold school district personnel accountable for "do[ing] better toward parents" (line 273). Mimi's statement prompts Lindsey to reflect upon the fact that "*all* of us have told about how in some way we helped somebody else" (lines 274–275), moving on to relate how she once challenged her son's elementary school principal on behalf of *all* children failing to read in his school (lines 275–281). However, the conversation returns to a focus upon changing students and parents (rather than the way we do education) when Kim introduces the idea of a class for "resource kids to teach them self-advocacy" (lines 285–286) with a simultaneous meeting for mothers to "find out ways we can support each other" (lines 290–291).

Katie engages the group in a discussion of teacher attitudes toward children labeled LD.

Katie: And I am *still*—and it just frustrates me beyond belief—that when I mention that my child is in resource, I get that automatic—"Oh gosh." The door is half open now. Whereas before—"Oh, yes! Let me have your child. We want your child!"
335 And then you go, "Well, my child is in resource. What do I need to do to make sure that these services are provided?" "Oh." I'm so tired of getting that *disappointment* response back from these people. I want somebody to openly be *excited* to have my child instead of that constant, "Oh, he's going to be a *burden*, but
340 we'll take him because we *have* to and it's by law that we have to. And legally we can't deny him, but we really don't *want* him." So then I'm suspicious.

Kim: Or are they really going to be *good* to him?

345 Katie: Or is this really the best place for him to be? So I think that maybe we should try to do what we can to lift the *burden*. Because obviously the funding is not there. So every teacher, you know, we gotta just face facts—it *is* a burden because they have to make certain accommodations. So I think that maybe
350 we should try to work even at the lower level—what can we as parents do to lighten the burden on the teachers? They have way too many students to be teaching anyway. So what can *we* do? So I was thinking maybe we should get the resource kids' parents or any volunteers within the school that your child happens to go
355 to and see if they, on a volunteer basis could come in and read tests to kids.

Kim: I tried to bring that up at my kids' school and it went over like a lead balloon. They did not want to. And that's a wonderful, open school with volunteers.

360 Mimi: I have a different take on that. I think that *far* too often—and you listen to the things we are saying in here—the *words* we are using

like—what can *we* do to *help* these teachers? But what *that* is, in my humble opinion, is that we come in *apologetic*. We come in defensive. We come in saying, "We are really sorry, but we are
365 bringing you *our* child. And what can *we* do to help *you*?" I mean you said it very well when you said you went to the teacher and said, "What can I do to help *you*?" And I was thinking then that you were almost apologizing. And *I'm* saying to you, being a mother, an aggressive, assertive mother when there was *nothing*
370 for children out there at *all* and from being a school board member, that the *last* thing that you need to do is to apologize. You've got to get an attitude change. You're going to have to come in and say, "I'm bringing you my child! And these are the things he is really good at." I don't even know how to develop that. When people
375 called me, I said, "First of all, I want to tell you that your child has to have an advocate." Children who are in the mainstream have *lots* of advocates. They've got that big cheering section like you were talking about. But these other children have *got* to have an advocate. Most of the time, they only have *one*, unfortunately,
380 and that's Mom.

Katie: *That's* why *I'm* saying—let's face reality. It's *not* going to change anytime soon either. So why *not* enforce some kind of program that will at least help these children until things can get changed?

Katie opens the conversation by expressing frustration about teachers' exclusionary attitudes, describing their disappointed response to her son's learning disability ("the door is half open now" [line 333]). Reflecting her previously articulated idea of a mostly empty cheering section for The LD Child, Katie laments, "I want somebody to openly be *excited* to have my child" (line 338). Moreover, she reports that teachers see her child's learning disability as a burden thrust upon them by law. Acknowledging that such attitudes make her "suspicious" (line 342), Katie reveals her sense of vulnerability to how teachers *choose* to regard and respond to her son, despite his legal right to a free and appropriate public education. Kim affirms her own sense of vulnerability to teacher attitude when she interjects, "Or are they really going to be *good* to him?" (lines 343–344).

It is of interest, then, that Katie engages with, rather than resists, the idea of "disability as burden" by suggesting that parents must "do what we can to *lift* the burden" of their children's educational needs (line 346). Revealing her acceptance of school as a place where some children belong and others do not, Katie asserts that "we gotta just face facts—it *is* a burden because [teachers] have to make certain accommodations" (lines 348–349). Rather than insisting that *all* children have the right to a free and appropriate public education, Katie suggests that the parents of "resource kids" (line 353) volunteer to assist teachers in making accommodations for their children (lines 354–356). It is worth noting that Katie assumes other parents to be in a position, like herself, to do so.

Kim considers Katie's proposal in light of her own attempt at proposing such a plan at her children's elementary school. She reports that "it went over like a lead balloon" (lines 357–358), providing no reason beyond "they did not want to" (line 358). At this point, Mimi abruptly shifts the dialogue in a different direction. She challenges the group to "listen to the things we are saying in here—the *words* we are using like—what can *we* do to *help* these teachers?" (lines 361–362), admonishing such a defensive and apologetic stance. Instead, she asserts that parents must advocate for their children's right to a free and appropriate public education and insist that teachers take responsibility for providing that education.

Katie uses Mimi's stated position to justify her own ("*that's* why *I'm* saying" [line 381]), arguing that parents must "face reality" (line 381) that teachers are not going to take responsibility anytime soon. It is worth noting that Mimi, at an earlier point in the conversation, makes a similar point when she states that "you can't *make* teachers do what they don't want to do" (lines 244–245), illustrating the kind of slippages that occur within our simultaneous engagement with and resistance against a dominant discourse (Haug et al., 1987). Katie asks the group, "So why *not* enforce some kind of program that will at least help these children until things can get changed?" (lines 382–383). Her insistence reflects a sense of urgency about proposing an immediate intervention that will benefit her child and others, rather than investing energy toward a long-range goal of holding teachers accountable under the law.

The group moves on to consider the general lack of teacher awareness regarding learning disabilities and the need for enhanced teacher education.

	Katie:	Nothing's going to change until the teachers that are being taught
630		to be teachers get taught to teach *all* children.
	Sue:	Well, maybe it's like they do in med school—where even if you
		are going to be a gynecologist, you still have to do a rotation and
		learn all these other disciplines.
		[group murmurs in support]
	Sue:	Because the thing that I've run into was the teachers don't know
635		*anything* about LD kids.
	Lindsey:	Nor do pediatricians.
	Sue:	They don't know *anything.*
		[overlapping talk among group]
	Kim:	And even some *resource* teachers who don't know.
	Sue:	They say, "What do you want me to do?" And I'm supposed to
640		tell them? They want *me* to tell them what to do about my child?
		Then I feel like I'm being a *bad mother* because I'm not an

educator! I'm an art director, you know! *You're* the one who is supposed to know!

Lindsey: Well, it's only been recently in the state that you could become
645 certified to teach *total* inclusion and *never* have any kind of course in special education.

Katie: *That* should be mandatory!

Lindsey: Now to get certified to teach, you have to take a survey class about the 12 different kinds of exceptionalities.

650 Sue: It seems pretty fundamental that teachers—I'm sorry, you can't graduate until you get your learning disabilities certification.

Lindsey: I tell my students all the time when they go out in the classroom—and that's one of the things I do is serve those student teachers—and I say, "You know, I know you're not going to hear me in the
655 beginning because you so focused on what you are doing. But this is *not* about *you*. This is about the children." I go sit in the back of a classroom when I go in to observe them and I watch for our little people. The ones who are like your son. And when I get ready to leave, I go out and I say to my student, "What about Susie?" "Oh!
660 She can't do *anything*." "Why?" "Well, her teacher last year said she can't do anything. She drives me *crazy*." I say, "Okay. When I come back next week, I want you to tell me *everything* that you can find out about this child. And I don't want to hear it from last
665 year's teachers." And when I come back, they're going, "I can't believe how smart she is. You won't believe! She doesn't even have a bed to sleep on at night." And then I get a whole new attitude from my students. And in *two* weeks, I go back and I got a whole new kid in that class. If we can just see *each* one as an
670 *individual*.

[group audibly agrees]

Lindsey: And I don't care if you have 500 of them. It's one at a time. And then you can respect them for *who* they are—and *not* see them for what they *can't* do, but for the *gifts* they have.

Drawing upon their lived experiences, the mothers concur that teachers appear unprepared to teach students labeled LD. Katie contends that teacher candidates must learn how to teach "*all* children" (line 630), revealing conceptualization of an inclusive classroom environment where her child *could* belong. Sue supports Katie's assertion, lamenting that teachers "want *me* to tell *them* what to do about my child" (line 640). Thus, the group recommends that all teacher candidates should be required to complete special education coursework.

Lindsey, who supervises teacher candidates in the field, suggests that teachers need not only information and instructional strategies, but, perhaps more importantly, the capacity to "see *each* [child] as an *individual*" (lines 669–670). Resisting the medicalizing of disability as pathology and deficit (Linton, 1998), Lindsey guides her teacher candidates to "respect [children] for *who* they are—and not see them for what they *can't* do, but for the *gifts* they have" (lines 672–673).

In the following excerpt, the mothers return to a discussion about holding teachers accountable for meeting their children's educational needs.

800 Mimi: What I'm saying is that I don't think that we should, as parents, say *anything* that gives the teacher *permission* to do any less for our child. We should not *pretend* to be understanding about that. This is the person who signed a contract to teach our children. Any time we give them permission, they *take* permission.
805 Kim: But how do you not do that and still have a good working relationship?
Mimi: I don't know **[voice lowers]**.
Sue: I think it's taking the approach that I am paying your salary. My child deserves the *same*.
810 Lindsey: I went on—be nice, be nice until you can't be nice anymore and then pull out your degree and then pull out the law. You work to be nice and have a working relationship until they take advantage of it, and then you say *no*.
Mimi: I think it's something about the attitude that you go in—that you are
815 proud of the child. I'm *proud* of my child. There's something that you can establish that makes that teacher *respect* you. But don't *ever* say anything that gives her the *slightest* bit of permission to slack your child.
[group murmurs as they consider what Mimi has said]
Mimi: Have you ever noticed in yourself that when your child's teacher
820 tells you something wonderful about your child that you have a different feeling for your child? Something elevates or something? Think about what that says. Society has taught us to step back and to not be assertive and proud and all these things about children who have a deficit. And that is proven by the fact that when somebody
825 tells you, like you said **[to Sue]** just then when your child was trying out and they said wow about his portfolio—you know, *you* had a different way of seeing your child in that he has esteem in *somebody else's* eyes and not just in yours.
Sue: Well, it *was* validation.
830 Mimi: And I think that's a *big* piece of it.

Asserting that parents should not "say *anything* that gives the teacher *permission* to do any less for our child" (lines 801–802), Mimi restates her belief that parents, without apology, must hold teachers accountable for providing a free and appropriate public education for all children. She points out that "this is the person who signed a contract to teach our children" (line 803)—a position affirmed by Sue who suggests taking the approach of "I am paying your salary. My child deserves the *same*" (lines 808–809). Yet, Kim questions how a parent can take a hard-line approach without alienating the teacher and sacrificing "a good working relationship" (lines 805–806)—a question for which Mimi admits she has no answer. In asking the question, Kim reveals her understanding of power relations that can and do influence the implementation of special education services.

Mimi and Lindsey reveal their distrust toward teachers, portraying them as likely to "*take* permission" (line 804) or "take advantage" (line 813). Lindsey describes how she "went on—be nice, be nice until you can't be nice anymore and then pull out your degree and then pull out the law" (lines 811–813); thus, Lindsey relies upon her cultural capital (i.e., a graduate degree in education, knowledge of the law, and her White, upper-middle-class status) to shift power relations in her favor (Brantlinger, 2003). Mimi, however, thinks beyond the cultural capital at her disposal, to consider that "there's something that you can establish that makes that teacher *respect* you" (lines 816–817). She goes on to challenge the group to think about how "society has taught us to step back and to not be assertive and proud and all these things about children who have a deficit" (lines 822–824). In making this statement, Mimi suggests that the origin of their challenges lies not only within the educational system itself, but also within larger *cultural meanings* of disability (Barton, 2004; Brantlinger, 2004; Linton, 1998; Reid & Valle, 2004; Ware, 2004). She muses, "And I think that is a *big* piece of it" (lines 830–831).

Conclusion

These five mothers, speaking across their respective generations, identify four themes common to their experiences within the special education system: exclusionary practices within schools, feelings of isolation, alienating professional practices, and professional indifference. Thus, it appears that these issues surfaced as the discursive practices of special education became established and have remained constant over time.

Given their experiences of marginalization, these mothers rather predictably offer suggestions for special education policy and practice that focus upon ways that parents and children can become more assertive *within* the present system. For example, they propose programs for the education of parents (support networks) and students (self-advocacy classes), leaving unchallenged the system that creates a need for such programs in the first place. As the conversation ensues, however, mothers consider the need for enhanced teacher awareness (regarding learning disabilities, specifically, and disabilities, in general) and teacher education (regarding instruction for students labeled LD). Only one mother, Mimi, suggests that the origin of their challenges lies not only within the educational system itself, but also within larger cultural meanings of disability.

Chapter Seven
Special Education as Ethical Practice

This four decade account documents the experiences of 15 mothers, representing diverse generations, social classes, and races/cultures, within the institution of special education. The narratives collected for this project illustrate *how* special education discourse—rules, structures, language—influences and shapes parent/professional relationships. In particular, these narratives consistently feature school professionals, operating from within an authoritarian discourse, who privilege objective ways of knowing (scientific, legal, bureaucratic) and dismiss subjective, contextualized knowledges that mothers bring about their own children. It appears, then, that these mothers of children labeled learning disabled (LD) (not unlike mothers of children with other disabilities) *clearly* represent the kind of "subjugated knowledges" that Michel Foucault (1980) defines as "a whole set of knowledges that have been disqualified as inadequate to their task or insufficiently elaborated: naïve knowledges located low down on the hierarchy, beneath the required level of cognition or scientificity" (p. 82).

What is striking about this collection of narratives is the emotional intensity with which mothers describe their exchanges with school professionals as well as their detailed retellings of incidents that may have taken place decades ago. Moreover, mothers reflect upon how such experiences continue to resonate within their lives. Thus, this collection of narratives represents what Foucault (1980) refers to as a *historical knowledge of struggles*, a particular kind of historical account that uncovers "the memory of hostile encounters which even up to this day have been confined to the margins of knowledge" (p. 83). It is my hope that this project contributes to an eventual centering of the margins of special education that will include the experiences and knowledge of mothers who navigate our system in pursuit of a free and appropriate public education for their children.

In mapping a history that documents mothers' experiences within *and* against special education discourse, I have attempted to meet Foucault's challenge by emancipating the narratives of 15 mothers whose experiences have been long suppressed by an authoritarian and scientific discourse—circulating and reinforcing itself—within the institutional machinery that drives special education. Although Foucault (1980) asserts that "power *is* always already there, that one is never 'outside' it" (p. 141, emphasis in original), he clarifies that this does not mean that "one is trapped and condemned to defeat no matter what" (p. 142). In other words, people always possess agency to resist domination. In fact, Foucault (1980) goes on to suggest that "there are no relations of power without resistances; the latter are all the more real and effective because they are formed right at the point where relations of power are exercised" (p. 142). Indeed, this collection of narratives attests to multiple points of resistance exercised by mothers across generation, social class, and race/culture.

Mapping Mother Narratives onto the Sanctioned History of Special Education

This alternative history of special education discourse begins in the 1960s, before the passage and implementation of the Education for All Handicapped Children Act (PL 94–142) within public schools. Although the field of learning disabilities emerges during the early 1960s, public schools do not institutionalize an educational response to children labeled LD until the passage of PL 94–142 in the mid-1970s. Public schools of the 1960s, however, do address the educational needs of some children with more visible and/or more significant disabilities within segregated spaces (separate classrooms and/or buildings). Other children simply are denied a public education on the basis of their disabilities, leaving parents to pay for private education and/or other disability services. It is worth noting that students identified as gifted and talented during the 1960s, however, receive "special" education within the public schools on the basis of their perceived potential to benefit American society. Thus, public schools of the 1960s operate out of the assumptions that (1) public education is for *some* but not *all* children and (2) it is natural and right to hierarchize and segregate children according to their perceived educational value.

The civil rights movement, a moment of significant rupture in American history that opens space for accommodating and distributing new discourses, includes a challenge to the rationale of educational segregation on the basis of race; yet segregation and/or complete exclusion of disabled children from public education remains the naturalized response to disability until the mid-1970s. It is worth noting, however, that the changes in American education brought about by the civil rights movement set the stage for the eventual emergence of a *particular* conceptualization of "special" education. For example, under the program outlined in President Johnson's War on Poverty, the federal government establishes Head Start and the Elementary and Secondary Act (ESEA) in an effort to improve educational outcomes for low-income students, thereby setting into motion a "discourse of cultural deprivation" (Sleeter, 1987; Spring, 1989). In other words, the cause of poverty is believed to originate in the disadvantaged lives of children rather than within the unequal social and economic systems of the United States (Brantlinger, 2003), neatly absolving public schools from the responsibility of school failure. Furthermore, the "discourse of cultural deprivation" includes the assumption that low-income children would be better prepared to enter school if their mothers were taught how to parent like White middle-class mothers (Spring, 1989).

The War on Poverty, with its emphasis on changing *parents* not schools, mirrors the paternalistic school rhetoric evident in the first half of the twentieth century (Valle & Reid, 2001). Of greater significance is the precedent for a government technology to channel federal funds for the purpose of identifying and serving a "special" Head Start population (meaning *deficient* as compared with the rest of the school population) who need "special" Head Start programs (meaning a *different* kind of education). Thus, the groundwork is laid for a subsequent discourse to materialize at the federal level regarding "special" education (meaning a different kind of education) for "special" children (i.e., disabled students who are deficient as compared with the rest of the school population). Moreover, it becomes a naturalized practice to construct and focus upon characteristics of "the individual" (i.e., the "culturally deprived child") rather than upon the interactions of "the individual in social context"—a practice that eventually undergirds the institution of special education.

What impact, if any, might these discourses have had upon mothers of children who struggled to learn during the 1960s? Della, a White middle-class mother, relates her experiences within an era in which neither special education services nor the conceptualization of a learning disability exist within public schools. In light of her daughter's early and persistent learning difficulties, Della repeatedly attempts to engage with school personnel; however, Della recalls that teachers treat Jennifer as if she simply were invisible within the classroom. Given that these federally supported educational programs exist only for "culturally deprived" students, school personnel regard Jennifer's educational needs as falling outside the realm of their responsibility (i.e., the school's responsibility is to educate those children who *can* learn). Thus, Della becomes positioned as the sole facilitator of her child's education. Unable to rely upon the public schools to educate her child, Della hires private tutors to supplement Jennifer's public education. It is noteworthy that it is considered right and natural that a

White middle-class parent *pay* to educate her child who is not learning within public school.

In the absence of any public school support for her child, Della, like other parents of financial means during the 1960s, turns to professionals within the fields of medicine and psychology for direction. Within a medical context, Jennifer becomes *highly* visible as the object of a professional gaze. Operating within a therapeutic culture that routinely medicalizes disability, these professionals rely upon the language of their respective fields, thereby framing Jennifer's learning difficulties in terms of pathology. Despite claims of rigorous scientific objectivity within their methods, it is noteworthy that Della recalls how different professionals simultaneously label Jennifer as both "normal" and "abnormal," yielding no valuable information, in either case, about how to teach Jennifer. Moreover, Della recalls that each of these experts conveys a similar message—that she, as the mother, must be responsible somehow for Jennifer's learning problems.

By the early 1970s, Cam and Mimi, both White upper-class mothers, begin their engagement with the public schools, and like Della, find teachers unable or unwilling to address their children's educational needs. They, too, rely upon their own financial resources to consult with professionals in medicine and psychology about their children. In response to an authoritative and scientific discourse used by such professionals, Cam and Mimi describe intense discomfort not only in response to the pathologizing of their children, but also to the pathologizing of themselves as mothers. Like Della, Cam and Mimi experience the "language of blame" used by professionals as deeply unsettling.

Meanwhile, on a national front, parents of children with disabilities, inspired by the civil rights movement, push for legislation that will guarantee the right of children with disabilities to a free and appropriate public education. Drawing upon this new and emerging social discourse in the early 1970s, Mimi challenges her local school district to pay her son's out-of-county tuition (to a public school for children with disabilities) on the basis that the district is unable to provide the education that her son requires. Mimi, a self-described trailblazer, successfully negotiates the tuition at time *before* PL 94–142; however, local school officials and professionals regard her as difficult and ultimately dangerous to their maintenance of the status quo. Certainly, Mimi does *not* behave as public school personnel have long expected mothers to behave—that is, obsequious and compliant in regard to school authority.

At this historical moment just prior to the 1975 passage of PL 94–142, it is relevant to consider that American public schools, of the mid-1960s to mid-1970s, have widely implemented "the open classroom." Based primarily upon Jean Piaget's groundbreaking child development research, the open classroom centers around the belief that "the learner *constructed* his or her own knowledge through the interplay of maturation, experience [both physical and cultural], and thinking about experience" (Reid, Hresko & Swanson, 1996, p. 224, emphasis in original). In contrast to traditional emphasis upon direct instruction and mechanistic learning, teachers in open classrooms facilitate instruction around active learning, interactions with others, and self-regulation.

By the mid-1970s, however, a Back to Basics discourse takes center stage for reasons that remain somewhat unclear (Cuban, 1993). Of particular significance is the notion embedded within the Back to Basics discourse that *some* people's children, unable to benefit from constructivist teaching in open classrooms, require highly structured instruction based upon the tenets of behaviorism (i.e., two types of children exist who need two distinct types of instruction). Thus, space opens to accommodate an impending notion that will support parallel systems of education—general education (influenced by child development theory and constructivism) and special education (influenced by behaviorism and technology).

The decade of the 1970s is an era in which the federal government becomes increasingly involved in legislating parents' rights. In 1974, the Family Educational Rights and Privacy Act passes into law, granting parents access to their children's school records and prohibiting the release of school information to a third party without parental consent (Cutler, 2000). The following year, Congress enacts the Education for all Handicapped Children Act (PL 94–142), representing the foremost legislation for children with disabilities and their parents in the history of American public schools. It is no longer deemed natural and right to exclude children from public education on the basis of their disabilities.

As this new discourse becomes embodied within the legislation of PL 94–142, the institution of special education emerges as the apparatus through which a "discourse of special education" becomes distributed. In other words, the institution of special education establishes "a system of ordered procedures for the production, regulation, distribution, circulation and operation of statements" (Foucault, 1980, p. 133)—all of which contribute to an agreed upon notion of Truth—thereby determining what *can* be said about children with disabilities. Moreover, children of average or higher cognition who struggle to learn become conceptualized as *being* "learning disabled," an identifiable "neurological condition" under the law that now necessitates "special" education.

Given that special education emerges from the disciplines of medicine and psychology, it is unsurprising that professionals readily adopt a medicalized discourse of disability along with a classification system dependent upon notions of normal and abnormal. Thus, a *particular* way of thinking about, talking about, and responding to children identified as "disabled" is set into motion. For example, a learning disability, now defined by federal guidelines as a discrepancy between IQ and achievement, requires *quantification* by its very definition. Given that IQ tests were developed within the scientific tradition, the legitimacy of such tools is not called into question nor is the practice of separating children on the basis of IQ scores. In fact, what naturalizes these ideas and gives them momentum is their association with the methods of natural science (Thomas & Loxley, 2001), thereby allowing space for the testing industry to arise within the newly established institution of special education. Therefore, it is considered right and natural for school psychologists, scientific tools in hand, to conduct the sorting of students into categories of normal and abnormal.

The introduction of the new special education discourse into public schooling, however, does not take place easily or swiftly. At this historical moment, arguments against integrating children with disabilities into public education bear striking similarity to arguments posited against Brown versus Board of Education and desegregation efforts during the 1950s and 1960s (Ferri & Connor, 2006). Special education is cast by critics as a burden thrust upon public schools (in much the same way that desegregation was cast as a burden) as well as a potential liability for students *without* disabilities—that is, children with disabilities might interfere with the education of students who *deserve* the opportunity to learn by virtue of their "normality." For example, Mimi recalls eagerly bringing her young son to his community school for the first time in his school career, only to bear witness to the principal's open disdain toward the integration of special education students within his elementary school.

Not only do children with disabilities gain unprecedented rights under the new law, so do their parents. Specifically, the law guarantees parents the right to be informed, the right to be knowledgeable about the actions to be taken, the right to participate, the right to challenge, and the right to appeal. However, as special education becomes institutionalized within public schools, these new discursive practices begin to function in a way that appears to *reinforce* professional knowledge—fortified by its grounding in *three* sources of expert knowledge (science, law, and education)—as knowledge exponentially superior to that which a parent brings about his or her own child. The experiences reported by this first generation of mothers under the new law (Cam, Mimi, Cosette, Lindsey) reflect an aggressive and authoritative professional discourse that positions them as passive participants in the process. Rather than feeling embraced as collaborative partners as the law intends, these mothers recall how legal, bureaucratic, and scientific language alienates them from the process of educational decision making. In particular, they recall school professionals who routinely privilege the "scientific language of testing" over the kind of subjective and contextualized knowledge that parents bring about their children.

In relating accounts of "shared" decision making, Cam relies upon assault imagery, while Mimi uses battle metaphors. Moreover, these mothers consistently report being perceived as irrational by virtue of their motherhood. In other words, school professionals, most of whom are women and mothers themselves, dismiss these mothers as too subjective, emotional, and enmeshed to contribute meaningfully—a decidedly patriarchal stance in conflict with the highly visible feminist discourse of the 1960s and 1970s. In fact, it appears that women/mother professionals who, in their *performance* of a masculine professional persona, set the terms of participation by privileging scientifically produced knowledge over other ways of knowing—paradoxically participating in the perpetuation of patriarchy (Lerner, 1986). For these women/mother professionals to believe what they are doing is in the best interest of others and to justify the dismissal of the mother's subjective perspective, they must choose to recognize themselves *only* as professionals and not *also* as women/mothers—a pattern that will persist across the decades.

In light of the historical tradition of professional domination over parents, it is somewhat unsurprising that school professionals regard the rights afforded to parents under the law as a significant anomaly (Skrtic, 1995) and rely upon familiar patterns of interaction. In response to professional authority, it is noteworthy that these mothers (White and middle to upper class) draw upon their cultural collateral (social class, material resources, male authority) to redistribute the balance of power and gain their rightful access to the process. For example, they access their material resources and social connections to elicit power (e.g., legislators, lawyers, university professors, private psychologists) that will lend credence to their point of view. Although the law guarantees parental participation, it appears that, *in actuality*, mothers must bring "authority" (most often male) to be able to participate.

By the mid-1980s, special education is established within public schools as *the* socially agreed upon institutional response toward educating American children with disabilities. In turn, school professionals (e.g., special educators, administrators, general education teachers, guidance counselors, school social workers, school psychologists) communicate in the "language of special education"—a *particular* way of talking about and responding to struggling learners—a hybrid language constructed from the merging of discourses (scientific, medical, legal, bureaucratic, and educational) that constitute special education practice. This new professional language becomes the taken-for-granted mode of communication for speaking about children with disabilities in public schools.

The second generation of mothers (Marie, Sue, Rosemary, Chloe, Dawn) attest to a persisting power/knowledge struggle between parents and professionals—not unlike that described by the first generation of mothers. Given that PL 94–142 (renamed in 1990 as the Individuals with Disabilities Education Act [IDEA]) represents a significant historical rupture in the way public schools must respond to children with disabilities and their parents, it is noteworthy that these mothers describe interactions with school professionals as authoritative and alienating, despite the law's *requirement* of parent/professional collaboration. It is striking, for example, that four of the five mothers (Marie, Rosemary, Chloe, Sue) who represent this generation (and identify as White and middle to upper class) exercise the legal right to secure an independent psycho-educational evaluation at their own expense in lieu of a free evaluation offered through the public school system. Citing distrust of school professionals, these mothers secure independent evaluations by third-party evaluators to ensure what they believe will be a fairer assessment as well as greater parental control over the knowledge presented and acted upon. In other words, these mothers do not trust school professionals to act in the best interest of their children, thereby reflecting a serious discrepancy between the kind of collaboration envisioned within the law and the *actual* interactions taking place between these mothers and school professionals.

Moreover, this generation of mothers regards "the language of special education" as a barrier to collaboration under the law. In other words, to participate meaningfully in educational decision making necessitates proficiency in negotiating a "discourse of law," a "discourse of science," *and* a "discourse of institutional implementation." Thus, Marie, Rosemary, Sue, and Chloe bring their third-party evaluators to special education committee meetings to act as mediators, a strategy intended to balance what they perceive as an inherent power distribution that favors school professionals. These mothers (like mothers of the previous generation) believe that school professionals regard their attempts to participate as somehow transgressing the rules rather than operating within them.

The second generation also reports that special education practice, with its emphasis upon the *individual* as the unit-of-analysis (rather than the individual within a *social context*), operates in a way that can construct a child as the sum of his or her deficits—a conceptualization often in conflict with a mother's contextual understanding of her child. Given that the discourse of special education reflects *particular* agreed upon assumptions about disability (i.e., disability as deficit or pathology), it follows, then, that children with disabilities *can only* be spoken about in this way.

Mothers of this generation report that school professionals regard the kind of contextualized and subjective knowledge parents possess as irrelevant to the agreed upon way of speaking about children with disabilities.

Furthermore, these mothers attribute the "deficit model of disability" to a growing ideology that two *types* of children exist—those who belong in general education and those who belong in special education—the latter becoming increasingly perceived by general education teachers as falling outside the realm of their responsibilities. For example, Lindsey, Cosette, Rosemary, and Sue document multiple ways that their children routinely become excluded *within* general education classrooms on the basis of their learning disability status. In other words, some general education teachers insist that children labeled LD perform exactly like the students *without* disabilities; if they cannot, they suffer the consequence of failing. It appears, then, that the introduction of a "place" (special education) within public schools for students labeled with disabilities leads general education teachers to conclude that such students *should* be taught by "special" teachers who teach "special" children in the "special places" designated for them. In fact, mothers consistently report that general education teachers regard the classroom accommodations and modifications on their children's individual education plans (IEPs) as a matter of teacher discretion rather than a legal requirement—as if the mandates of special education do not apply to teachers in general education.

It is relevant to consider that during this historical moment (mid-1980s to mid-1990s), mounting evidence supports the mothers' observations of a growing ideology among educators that two *types* of children exist—those who belong in general education and those who belong in special education. Of particular concern among emerging critics of special education is the overrepresentation of African American and Latino children in segregated special education classes (Brantlinger, 1997; Lipsky & Gartner, 1996; Skrtic, 1991, 1995). Such overrepresentation suggests that school personnel perceive African American and Latino children as *needing* special education more often and to a more significant degree than White children. Furthermore, given the long tradition of American schools as a vehicle through which to promote the majority culture's political, social, and economic agendas, it is unsurprising that general education teachers begin to refer immigrant students (who arrived in the United States during the influx of immigration during the 1980s and 1990s) to special education because of mistaken assumptions about their abilities (Lagrander & Reid, 2000).

Despite the guarantee of a free and appropriate public education for their children within a system in place now for one to two decades, it is noteworthy that *all* of the mothers in the second generation rely upon private resources to supplement their children's "special" education (e.g., tutors, educational consultants, psychologists). It is of particular interest that their narratives bear striking similarity to Della's account, which takes place during the 1960s *before* the law's implementation.

In light of persisting tensions between parents and professionals in regard to shared decision making and IEP implementation, mothers of the second generation, like the previous generation (White and middle to upper class), draw upon their cultural collateral (social class, material resources, and male authority) to gain rightful access to the process. Despite bringing such "power" to the table, these mothers, like those of the previous generation, describe interactions with professionals in adversarial terms. For example, Marie refers to special education committee meetings as "a big battle" between opposing teams, while Rosemary describes school professionals as fighting her "tooth and nail" and "dig[ging] in their heels." Chloe's use of assault imagery mirrors Cam's description of events that occurred nearly two decades earlier.

Moreover, this generation continues to document that professionals hold biased assumptions about their rationality based upon their embodiment as mothers. This point is particularly illustrated within Marie's narrative. As a professional in the learning disabilities field *and* a mother of a child labeled LD, Marie consciously chooses to perform the role of professional rather than the role of mother. It appears Marie is unable to conceptualize a space in which she can be mother *and* professional. As an insider to the profession, Marie recognizes the weak position of mothers and strategizes accordingly.

In the most recent historical era, defined within this study as the mid-1990s to 2004, special education is firmly entrenched within public schools as the naturalized response to children with disabilities. However, a growing number of critics question the efficacy and ethics of what has become a separate and largely unequal system of education (e.g., Allan, 1999; Ballard & McDonald, 1999; Brantlinger, 1997; Lipsky & Gartner, 1996, 1997; Sapon-Shevin, 1996). In response to evidence of special education students routinely being excluded from the general education curriculum, the law's 1997 reauthorization requires that special education students have access to the general education curriculum. (It should be noted that this IDEA amendment validates the experiences reported by the second generation of mothers in this study.) Thus, competing discourses circulate among researchers, policymakers, school professionals, parents, and administrators regarding the best way to meet the educational needs of special education students—segregated in "specialized" settings with other students with disabilities or included in general education among students with and without disabilities.

It is appropriate to situate the inclusion debates within this most current era in which a "discourse of standards" and a "discourse of testing" permeate American public schools. It appears that members of the majority culture intend to save our schools by imposing higher educational standards in the name of equalizing opportunity for all—a political stance particularly well reflected within the rhetoric of the *No Child Left Behind Act* (2001). Thus, while proponents of inclusion promote classroom community, social learning, and difference as a resource, they do so within a school culture that is increasingly fixated upon academic standards and standardized assessment.

The most current generation of mothers (Elsie, Kanene, Kim, Katie, Debra) describes multiple ways in which their children experience *exclusion*, despite increasing acceptance of more inclusive practices within public education. For example, Elsie and Kanene, describe how their children, saddled by low teacher expectations for students with learning disabilities, become excluded by virtue of their imposed invisibility within the classroom. (It is worth noting that Elsie and Kanene's experiences closely reflect narratives told by the first generation of mothers.) On the other hand, Kim, Katie, and Debra report that their children, expected by teachers to behave exactly like children without disabilities (i.e., the children who *really* belong), become highly visible in classrooms where they fail to meet these expectations. Debra, in particular, observes the irony of IDEA guidelines that guarantee an IEP on the basis of her child's specific educational needs as a student labeled LD while simultaneously her child must meet the educational standards expected of *all* students under *No Child Left Behind* (2001). She questions the logic that holds her son accountable to educational standards without consideration of his *school*-documented learning disability.

Much like the narratives told by the second generation of mothers, these mothers also experience general education teachers who perceive special education students as a burden thrust upon them rather than their responsibility under the law. No doubt the intensifying pressure of academic standards and the significance placed upon standardized test scores contribute to teachers' perceptions of special education students as a burden in the general education classroom. Mothers who insist that teachers assume their responsibility become constructed not only as demanding but also as overinvolved with their children. It is of interest that fathers who insist do not become constructed in the same way. Thus, mothers of this generation also contend with *gender bias* by virtue of their motherhood—a pattern well established across the decades.

Narratives of the present generation suggest that mothers continue to be passively positioned by an authoritative discourse that claims the power to define, classify, and limit children. Echoing mothers of the previous two generations, these mothers describe their interactions with school professionals in adversarial terms. For example, Elsie likens special education committee meetings to The Inquisition, while Chloe and Kanene rely upon words that connote assault and aggression to describe their experiences. These mothers actively resist routine objectification of their children—for example, Elsie brings a photograph of her child to a special education committee meeting, Kim dismisses as irrelevant "those numbers" that describe her child, and Kanene refuses additional testing by asserting that her son is a human being and not "a pincushion."

Elsie and Kanene, who represent nondominant culture (mothers of color who self-identify at or below poverty level), observe that dominant culture mothers (White and middle to upper class) are better able to command attention from school professionals because of their inherent cultural collateral. Indeed, Kim, Katie, and Debra, as members of the dominant culture, acknowledge drawing upon the resources available to them (e.g., White privilege, social class, material resources, and male authority) in order to gain greater access to the process. Moreover, Kim and Katie, like the White middle- to upper-class mothers of the past two generations, rely upon private resources to supplement their children's "special" education (e.g., tutors, educational consultants).

In conclusion, this collective narrative told across four decades represents an alternative account to the official rendering documented within professional annals of special education. It is worth noting that mothers of each generation describe themselves as undergoing a personal, and often painful, evolution as they learn to negotiate within the complex and bureaucratic institution of special education—a system that presumably exists for the *benefit* of parents and their children with disabilities. Each mother tells a story of her particular journey from a passive participant who places trust in experts who speak an authoritative discourse to an active participant who advocates for her child's rightful education. It is of particular interest that each mother identifies a specific moment within her narrative that represents a "turning point" after which she refuses to be passively positioned again.

What is perhaps most striking is the constancy of issues (e.g., alienating and authoritative professional language, conflicts in "shared" decision making, exclusionary classroom practices, professional disregard for other "ways of knowing") that mothers report across time. Furthermore, these issues exist, regardless of social class or race/culture, the regional area of the country where mothers reside, and whether or not their children attend public school in urban, suburban, or rural contexts. It is striking that Della's narrative, a story that takes place *before* the institutionalization of special education within public schools, is retold many times over in the narratives of all the mothers who follow. It is highly significant, then, that these three particular generations of mothers with children labeled LD report that this landmark legislation, in *actuality*, has engendered *minimal* impact upon the quality of their children's education.

Consequences of the Discursive Practices of Special Education

My intent is neither to dismiss the existence of any positive effects of special education nor to diminish any successful experiences of parent/professional collaboration that other mothers may have had. Rather, I hope to have brought clarity to the *process* by which special education and its discursive practices *can* operate in ways that produce negative consequences for mothers and their children. I argue that a learning disability can be understood as a socially constructed discourse, within a particular historical and culturally situated moment, that operates in agreed upon ways in response to children who experience difficulty learning. (Such a conceptualization does not, however, preclude the biological basis for learning disabilities.) Thus, this project contributes to our understanding of *how* disability becomes socially constructed, specifically within the context of school. Of greater significance, the project documents mothers' perspectives on the consequences of special education's discursive practices within their lives.

By superimposing narratives of 15 mothers, spanning four decades, against the sanctioned history of special education, I attempt Foucault's goal of posing a philosophical challenge to history—"not to question the reality of 'the past' but to interrogate the rationality of the 'present'" (Gordon, 1980, p. 242). In light of this 40 year history as told by mothers, it appears germane indeed to this work that we consider questioning the rationality of our current system of special education.

In addition to the consequences of special education practice well documented by the mothers in chapters three–five, I consider it relevant to the goal of "questioning the rationality of the present" to include what mothers also say about the impact of special education discourse upon their self-image as mothers, their children labeled LD, and their families. Their stories are grouped into three broad themes: (1) consequences for self-as-mother; (2) consequences for the child labeled LD; and (3) consequences for the family.

Consequences for Self-as-Mother

As well documented within chapters three–five, mothers describe feelings of grief, fear, anger, shame, hopelessness, and guilt in response to an authoritative scientific discourse that constructs their children in terms of pathology. Across generation, social class, and race/culture, mothers repeatedly explain how they feel targeted as objects of blame by this authoritative discourse. Cam, for example, recalls how she saw herself as an inadequate mother.

C: The first
700 stages, as a mother, you're trying to take care of yourself emotionally in trying to deal with a child that maybe reflects on you not being as *good* as you should be to society. It's not that you *feel* that way but that's how society looks at you. You did something wrong. You have a handicapped child. You're grieving and dealing with *all* these labels
705 and all of these *guilt* feelings—right or wrong—that you have inside.

Feelings of inadequacy become so great for some mothers that they not only question their competency to parent, but also their purpose in life. For example, Cosette, a mother of three, explains the depth of her emotional response to interactions with authoritative professionals.

C: And I remember thinking so very clearly that I certainly do not need to be the mother of three children. If I had this child who had this
240 kind of behavior and I've been told that one—it was brain damage, and I was told by the pediatrician that there wasn't anything wrong with my child, and then I go to this child psychologist who was supposed to be so well respected and I'm told that all the problems are *me*—then I don't really need to be having three children or I don't need to be in this
245 position on the earth. Slowly over the next four or five months—I became *very suicidal*. And thought, I just don't need to be a mother, somebody else could mother these children better than I could [**voice lowers**]. That really did happen [**deep pause**].

Cam admits similar feelings after university experts advise her to institutionalize her elementary-age daughter.

C: I remember flying home that night. Pat, Bob, and I were flying home from Memphis that night and
185 I was *completely devastated*. And we were flying through a thunderstorm. You could see the lightning out the window of the plane. I remember thinking, and I'm not a person who has negative thoughts, but I remember thinking, "If this plane goes down, it's all right because I *cannot do* what they're telling me to do. If that's the way I have to live my life, I *simply*
190 cannot do it."

The effects of an authoritative professional discourse that negatively impact a mother's self-image may continue to reverberate within her life for *years* afterward. For example, Della has difficulty getting her words out when speaking about her exchanges with professionals that took place more than *three decades* ago, revealing the significance of emotion that remains for her as mother.

D: It is frustrating [**long pause**]. It makes you feel like a *failure* [**pause**]. Not a good feeling [**voice breaks**].

Mothers from each generation attest that the scientific discourse that deems their children outside the realm of "normal" circulates well beyond the school context to influence relationships in other social settings. For example, Cosette speaks of her sense of isolation from mothers of "normal" children who seem unable to understand her life and children.

C: I think that it is very, very hard as a mother with special needs children to have lady friends who have "normal" children. You just can't be really
435 close because you have so many frustrations with your LD kids and stuff that your other lady friends just cannot understand. There is *no way* for them to understand what you live with on a daily basis. So, you wind up gravitating toward other mothers or finding other people who perhaps might understand where you're coming from because
440 a regular mother with "normal" children—she's clueless. I could never be friends with a mother who had normal children. It just didn't work.

Similarly, Katie relates how mothers of children without disabilities frame her child's learning disability in terms of pathology and respond as if it is something too terrible to talk about.

K: When I mention it to mothers who don't have children with learning disabilities, it's
kinda like, "Oh." They feel uncomfortable. It's almost like saying that my child has a fatal disease. And they go, "Ooooh." It's a little pity sound in
550 their voices. And they quick try to change the subject.

Given that a disability discourse rooted in deficit and pathology circulates within social contexts other than school, Cosette explains how exclusion extends beyond her children to herself as a mother.

C: I think, for the most part, you get the message that it must be a discipline problem in the home if your child acts in any of these ways—that it gets this particular kind of label. You always feel that they think there's
405 something wrong with *you* and normally these children have social issues. Their social skills are usually very weak. These children usually

don't have very many friends. And a lot of times when you have young children, *your* friends are often the result of parents of children of your children and we never had the pleasure of that. My children—
410 two of my three children—were very poor at social skills and didn't really have any friends when they were quite young. They would never get invited to birthday parties. If they were to have a birthday party, kids would probably would not come. That happened on numerous occasions. And that wound up being *extremely* painful for a *mother*—
415 unbelievable—and also for a child. And most of the times, you just kind of feel like a *second-class citizen* or an *outcast.*

Cam shares similar experiences as the mother of a child labeled with a disability and explains how she learns to resist cultural assumptions about disability thrust upon her child, herself, and her family.

C: One of the challenges is that society takes all these terms that we put on children and they have *attitudes* about those children. And I had to learn that it didn't matter. I had to *honestly* learn within myself that I couldn't
495 change what society thought. And I'm talking about other dentists' wives or I'm talking about the Sunday school teacher. I'm talking about the friend next door. I couldn't *control* what they thought other than to try to educate them. If people thought there was something wrong with us because we had a handicapped child, then that was *their* problem not our
500 problem. There was nothing *wrong* with us. I had to *get over* the embarrassment of having a learning disabled child. And that—*she* taught me well.

These narratives suggest that a medicalized conceptualization of disability can engender feelings of intense inadequacy and guilt in mothers. Moreover, mothers document how the assumption of disability as deficit, circulating within the larger social context, not only positions their children as outside the realm of "normal," but also themselves—thereby negatively impacting their relationships with mothers of so-called normal children. Thus, it appears that the conceptualization of disability as pathology impacts a mother's self-image as well as her perceived desirability in the eyes of other mothers.

Consequences for Child Labeled LD

Every mother in the study, regardless of generation, social class, or race/culture, speaks to the enormous time investment needed, for both parent *and* child, to meet the requirements of the school curriculum. Their narratives illustrate the intimate, day-to-day lives of children who struggle to keep pace with their peers. For example, Della, speaking from an era before the passage of PL 94–142, recalls teaching her first grade daughter every single evening.

> D: The teacher brought her homework out to the car and gave it to me. She explained it to me so I could help Jennifer with it. Then when Jennifer got home, we would sit at the kitchen table and we would struggle the rest of the afternoon. It took us until dark. So that meant she had no time outside with her friends—[**voice lowers**]—what friends she had—or at least, she couldn't go out much. Most every day, we struggled.

It is striking that Katie and Kim, speaking from the present and with the benefit of special education services for their sons, respectively, mirror Della's experience of nearly 40 years earlier.

> K: So this kid, other than the days he would have practice, was doing *work* nonstop. He wasn't going out and playing as a normal child. To me, that child had not a normal childhood. I felt like he was being *robbed* of his childhood because of this learning problem. We just
> 205 couldn't seem to get him in the right direction—or *fast* enough to meet the demands that the school was asking of him.
> K: And I was in tears because I couldn't take care of anybody else. I couldn't get dinner on the table. I couldn't spend any time with my other kids because I had been with him for *hours*. And *that* was
> 110 pretty much our daily *grind* after school. We would sit at the kitchen table and we would watch one by one the other kids finishing their homework and going out to play in the neighborhood. And all the neighborhood kids would be in the cul-de-sac and he'd be at the kitchen table, rubbing holes with an eraser in his paper. And the most frustrated he would ever get
> 115 would be—he'd wad up his paper and start again, but he never gave up. And I would say, "You know what. Let's just put this away until dinner." And he wouldn't. He was determined to get it finished—to the point that it was way too long.

Elsie not only teaches her child at home, but also becomes a full-time volunteer in her daughter's inclusive elementary classroom so that she can better monitor and strengthen Emily's education.

> E: And everything was fine because I was there and I really took on a lot. The teacher taught. I learned what she taught and
> 505 taught it to Emily after school. So we had two years of this. Coming home everyday, blah, blah, blah. Hitting the books. Teaching it all over again because I had to teach her at a slower pace.

Rosemary describes the endlessness of homework support from year to year as a "merry-go-round [that] just keeps going and going. It doesn't stop. You never get off...you just go round and round." Sue likens daily homework completion to "another job—another gut-wrenching job." Debra, on the other hand, describes anger and frustration in regard to homework expectations, lamenting what she perceives as a lack of support on the part of her son's special education teachers.

D: You become very angry that they don't have time for your child [**deep pause**]. I'm looking at Justin as his *whole life*. As a mother, you're
330 responsible for that child's development for their *entire* life, not just for this school year. The teachers look at it just for the school year. You feel like when you're home being a mom when you're working full time, there's just so much time in a day. You want to give as much time as you can to your child. But you also have another child that needs your time.
335 They really do put everything on the parent to make sure all the homework is correct and everything is done. They don't want to help him with any homework or anything in school. It's all up to the parent. It's just *frustrating.* I don't know how else to put it. It's just *total* frustration. These resource teachers went to school to be a resource teacher and to help the
340 child and all you get back is—well, no you're the parent, you need to do it. And I feel like they're just there filling out the needed paperwork. That's all they do. They go in. They fill out the needed paperwork that the law requires. They sign it off and their job is done.

Furthermore, mothers painfully acknowledge that even hard work does not guarantee success for their children. Lindsey, for example, relates her anguish and helplessness as she watches her young son lose his enthusiasm for learning.

L: He was trying and he was doing the work—like those darn hundred problems of math that he was doing wrong. He'd do every single
280 hundred one of them she'd ask him to do. And so it was not that he had given up, but he was starting to give up. I often describe that like a candle that's being blown out and you just watch it, slowly slowly slowly suffocated and going out. You could just see that in him.

Throughout this collection of narratives, mothers also speak to the negative consequences of labeling in their children's lives. They report that children without disabilities, picking up and acting upon cues from the social context, routinely exclude, bully, and/or ridicule their peers who bear labels. Della, Sue, and Katie, who represent each generation of mothers, speak respectively to the consequences of labeling in the lives of their children—suggesting the ongoing nature of this issue.

D: [**long pause**] But, she was, I know, being laughed at and ridiculed and bullied and made fun of.

I wish I had known back then what I know now. And I think her emotional and social life might be different today [**voice breaks**].

S: I got a
225 call that some bully in the science classroom, in his advanced science class, had targeted him and was harassing him every day. The teacher left the room for five minutes. The kid came down the aisle, threw some paper in front of Evan and told him to pick it up. And he refused to pick it up. He pulled him out of the desk by his shirt and he and three other boys took
230 him and stuffed him in the trashcan head first. On another occasion, he was choked from behind by a belt, by the neck. And this is all the stuff that this kid had to put up with.

K: He still has to deal with the insults. He still has to deal with the name-calling. I'm not saying it happens on a daily basis, but from time to time. And he's so sensitive to it. *It's* very *painful.* It hurts.

Such stories suggests that a "special" education alleviates neither academic nor social stressors for children labeled LD. In fact, mothers document intense stress in regard to teaching their children at home (to the exclusion of other recreational or family activities) in an effort to keep them afloat. Thus, "special" education does not appear to provide the educational support necessary for these children to work independently and efficiently at home. Moreover, mothers report that their children routinely become positioned by peers as "low status" because of their disability labels.

Consequences for Family

In light of the tremendous investment of time required to monitor and facilitate the education of their children labeled as LD, mothers speak to the impact of this situation upon their other children. For example, Kim expresses concern about future consequences about which she may not even yet be aware.

K: I'd also like to make a comment quickly about how it's impacted my other children. It has *definitely* impacted them. So many times, they've come to me and said, "I'm supposed to read this out loud to you (this is my second grader—
765 recently—last year) and you're supposed to sign it." And I find myself saying, "Just read it to yourself and I'll sign it. I'm working with Jacob and I don't have time for you to read it to me. I *know* you can read it, honey, and I *know* you're a great reader. Just go do it." I fear that my other kids will turn it around and say, "Jacob needs Mom's help all the time
770 and we don't get any." It's really not been a problem, but I've made sure that—well, yeah, it probably has been a problem—it's probably affected them in ways I don't even know [**voice lowers**]. And I've probably just hoped that they weren't being affected by it.

Mothers further suggest that this intense investment of time with the child labeled LD impacts not only other children in the family but their marriages as well. The following excerpts from Elsie and Rosemary, respectively, illustrate this point.

E: Parenting a child with LD [learning disability] to this degree because I really went in it into a full throttle—I say that part of the reason that my marriage suffered and I'm divorced—part of it was because of that. My focus was always
1355 on children. I know they say your husband should come first and stuff, but for me, even with my biological children, children first for me *always*. I think it may have had an impact on my marriage.

R: I think that my involvement with it is *one* of the things that led to our divorce. Because he began to get attention from *other* women. I was
750 busy taking her to tutors, tutoring her—you're *exhausted*. When you're working with a child like this trying to coach them, you become mentally exhausted. And, no it did not *cause* the divorce. I'm not trying to blame that. I *really* am not. But that's a good excuse. Because when you're "making it" out in the world and then there are
755 all these women out there and they're patting him on the back and oh, that's *wonderful* and with me, I'd say, "Oh thank you. Oh, that's good. Oh, I'm glad you won that case. But I've got to get over here. We've got to finish these two chapters tonight." It *never* ended! I do think that that had *some* effect. Nobody has ever told me that, no
760 counselor has ever said that. That is my *own* gut feeling. I would not *ever* want Rebecca to think that caused it. So, I do not think that did it. But I do think it contributed to the divorce [**voice lowers**].

And as Debra notes, even for marriages that remain intact, stress is undeniable.

D: It is very stressful on a marriage because it takes so much extra work. It takes extra
440 work from your marriage. There is less time for a couple. I mean there's less time anyway when you have children. There's very little time to be a couple. But when you have one that needs *so much* extra work, so much extra attention, it's very difficult on staying a couple and finding time to be that couple. It's hard. But again, you have to also agree that as a
445 married couple that this is going to be your focus and deal with it.

Beyond the stressors present within their own families, mothers identify extended family members as yet another source of tension regarding their children labeled LD. In the following excerpts, Marie, Chloe, and Katie, respectively, address how such tensions play out within family life.

M: My mother-in-law was *horrified* to hear that we had *deficient* children—as she referred to

them. And when I said, "No, they're *not* deficient. They just learn differently," she said that she taught kindergarten for 30 years and she
1100 *never* had a child like either of them.

P: My husband's family, on the other side, were people who *never* believed. They wanted *nothing* in the family that even *remotely* touched on a mental condition. "We don't have this in
595 our family. If we do, we don't let *anybody* know what's going on." They *never* accepted the idea of learning disabilities.

K: I was pretty much *shocked* once we had Louis diagnosed—the misconceptions out there [**pause**]. Especially with my *own* mother. When I told my mother that Louis got diagnosed with ADD and with a learning disability, she
655 could accept the learning disability but not the ADD. She said, "Oh, that's ridiculous. There's too much of that going on and I think parents are just not willing to be parents and correct bad behavior and too many kids are on medicine just because the parents are lazy [**laughter**]." So, I was going, "Ahhhh, no." I just chose to ignore her.

These narratives illustrate how the conceptualization of disability as pathology circulates well beyond the school context to impact not only the child labeled as LD but immediate and extended family members as well. In light of the tremendous investment of time committed to educating their children labeled LD, these mothers acknowledge that relationships with their other children suffer as well as their marriages. It is clear that our current conceptualization of disability as pathology can result in highly problematic and long-lasting consequences for families.

Conclusion

Speaking for their children who cannot speak for themselves, mothers, in effect, are the epicenter through which all disability discourses pass. As such, mothers continually respond to multiple and competing sources who claim knowledge about their children (e.g., psychologists, teachers, relatives, acquaintances, friends, doctors, even strangers). It appears, then, that mothers serve as human repositories for cultural assumptions regarding disabilities in general and learning disabilities in particular. It is worth noting that these mothers experience this barrage of opinions (solicited and unsolicited) as uncomfortable, confusing, and often painful, struggling to articulate and/or sort out the source of their emotional responses.

Furthermore, it appears that the discursive practices of special education (language and structures) impact children labeled LD as well as their families. These mothers speak about the consequences for themselves as mothers (e.g., low self-esteem, depression, difficulty relating to mothers of "normal" children), their children labeled LD (e.g., social exclusion, stigma), and their families (e.g., hyperfocus on the child labeled LD, divorce, resistance of the disability label by extended family). Thus, the effects of our particular disability discourse reverberate far beyond the school context and into the everyday lives of children and their families.

It is my hope that this type of narrative research will continue to gain greater presence and credibility within the field of special education. I contend that the rather static nature of the field (documented by the constancy of issues that mothers report regardless of generation, social class, race/culture, geographical area, or school context) exists, in part, because of our field's preoccupation with research and practice primarily grounded within science. I am not disputing the value of scientific research, only the tradition of privileging it. I envision a broader space in which narrative research is embraced alongside more traditional research in special education.

I believe that the emerging discipline of disability studies in education holds great promise as a framework for conducting special education research. Rather than defend the present system of special education, we must be bold in generating critical questions about the *consequences* of special education practices upon those intended to "benefit." I believe that we must begin to consider our conceptualization of disability and the structures that support it in terms of *ethics* rather than science.

I hope to continue a line of narrative research with fathers, guardians, and sets of parents/partners. Given the rather limited data that I was able to collect from mothers who represent nondominant culture, I am interested in further exploring how these mothers experience the institution of special education. I believe that we can be critically reflective about our field *only* by engaging with our "stakeholders" (i.e., students labeled as disabled and their parents) to identify problems and contribute to solutions.

It appears germane to my line of narrative research to investigate how teachers, working in both general and special education, experience the institution of special education and their interactions with parents of children labeled LD. Given the negative manner in which mothers in this study describe many teachers, it would be of particular interest to understand the process by which teachers likewise become positioned and constructed by the discursive practices of special education. In other words, how are teachers constrained by the available discourses? In what ways might teachers resist being positioned by special education discourse? Moreover, I would like to solicit the perspectives and experiences of school professionals *and* parents collectively. I believe that focus groups as well as in-depth and extensive group interviews, in particular, could be invaluable in contributing to our knowledge base about parent/professional collaboration.

A Message to the Profession

In constructing a collective narrative from the perspective of mothers, I focused upon identifying points of connections and differences among three generational groups of mothers whose narratives span four decades. However, I wish to emphasize that the narratives collected for this study represent the *particularities* of individual human lives impacted by *particular* experiences within our system of special education. I believe that it is upon these particularities that we must focus our attention.

I urge us as professionals to focus less on our own agendas and procedures and far more upon the everyday lives of children and families. As the narratives documented within this study illustrate, the experience of parenting a child with a disability can be wrought with multiple and sometimes conflicting emotions. It is of interest, for example, that mothers vehemently advocate for their children's rightful place in the world "as they are" while simultaneously revealing a desperate desire for their children to be like everyone else. Several narratives illustrate how mothers, on the one hand, argue passionately against the hierarchal nature of public schooling, yet on the other hand, clearly differentiate their children labeled LD from children labeled "retarded" in a similarly superior manner, thereby engaging in, rather than resisting, the hierarchizing of children according to ability. The roads these mothers travel are treacherous indeed. I believe that we can begin to move closer in partnership by opening a space in which to *really* listen to the particular stories that mothers need to tell about their children and themselves. I remain struck by the fact that nearly every mother in this study described the interview process in transformative terms, citing the meaning making that occurred in the telling of her story. Moreover, a number of mothers indicated how grateful they are to have the audiotapes and transcriptions as a record of their family's story.

I have long been concerned about the consequences of special education practices upon the lives of mothers and children. Over the years, I have been witness to countless school professionals, mired in the habituated practices of special education, who contribute to unintended negative consequences by failing to attend to the particularities of mothers and children—and remain unaware that such consequences have even occurred. Thus, it is my hope that this kind of narrative work contributes to a growing body of research that will inform the profession about the perspective of mothers and enhance attention to the particularities of families.

If we begin to focus more upon *imagining* possibilities for mothers and their children than upon our habituated procedures, we might move closer to the kind of parent/professional collaboration envisioned by the law. Maxine Greene (1995), a renowned philosopher of education, explains that

> imagination is what makes empathy possible. It is what enables us to cross the empty spaces between ourselves and those we teachers have called "other" over the years. If those others are willing to give us clues, we can look in some manner through strangers' eyes and hear through their ears. (1995, p. 3)

Evidence of our conscious or unconscious oppression of mothers lies within their resistance to our practices—for there is no resistance without the presence of power. As revealed within the narratives of mothers, resistance takes place in multiple forms at the local level where power is exercised. We must pay attention to the "special cases" and their points of resistance—for it is within such resistances that we see the unintended consequences of our practices.

If we are ever to achieve the kind of collaboration with parents envisioned under the law, we must be willing to let go of our expert stance to consider other equally valid "ways of knowing." I contend that we can move closer to parents only by embracing their subjectivity.

> Each of us achieved contact with the world from a particular vantage point, in terms of a particular biography. All of this underlies our present perspectives and affects the way we look at things and talk about things and structure our realities. (Greene, 1978, p. 2)

By imposing our objective realities upon parents while dismissing their subjective realities, we perform the dangerous act of disqualifying another's knowledge which, as philosopher Richard Rorty (1989) cautions, can "make [one] incapable of having a self because [one] becomes incapable of weaving a coherent web of belief" (p. 178). The consequences that result from the act of dismissing a mother's reality are significant and well documented within the pages of this book. We can no longer ignore the consequences of our profession upon the lives of families. We must hold ourselves accountable for the *ethics* of our practice.

Furthermore, the quality of our professional work becomes significantly diminished by insistence upon a single "way of knowing." Let us heed Heshusius' (1994) call for a participatory mode of consciousness within the field of special education in order to develop

> the awareness of a deeper level of kinship between the knower and the known. An inner desire to let go of perceived boundaries that constitute

> "self"—and that construct the perception of distance between self and other—must be present before a participatory mode of consciousness can be present…it involves letting go of the idea of being-separate-and-in-charge altogether. (pp. 17–18)

As special education professionals, we must challenge ourselves to think beyond compliance toward imaginative dialogue with parents. We would do well to embrace Greene's (1995) conceptualization of the kind of thinking that

> refuses mere compliance, that looks down roads not yet taken to the shapes of more fulfilling social order, to more vibrant ways of being in the world. This kind of reshaping imagination may be released through many sorts of dialogue. (Greene, 1995, p. 5)

My hope for the profession is that we develop the habituation to "look at things as if they could be otherwise…to summon up the 'as if,' the might be, that which is not yet" (Greene, 1998, p. 1)—to imagine new ways of conceptualizing, talking about, and responding to disability that refuse the pathologizing of children and their families and open space for authentic partnerships to flourish.

Appendices

Appendix A: Participants and Settings

Participant	*Southeastern United States*	*Northeastern United States*
Della	small town	
Cam	medium-sized city	
Mimi	small town	
Lindsey	small town	
Cosette	medium-sized city	
Rosemary	medium-sized city	
Marie		small suburban town
Chloe	medium-sized city	
Sue	medium-sized city	
Dawn	rural area	
Elsie		large metropolitan city
Kim	medium-sized city	
Kanene		large metropolitan city
Katie	small town	
Debra	small coastal town	

Appendix B: Generational Groups of Participants and Children

Early Years 1960s to mid-1980s	*Implementation Years mid-1980s to mid-1990s*	*Maintenance Years mid-1990s to 2004*
Della Jennifer (b.1960)	**Rosemary** Rebecca (b.1981)	**Elsie** Emily (b.1988)
Cam Pat (b.1967)	**Marie** Tia (b.1979) Gregory (b.1982)	**Kim** Jacob (b.1989)
Mimi Pete (b.1968)	**Chloe** Jessie (b.1982) Ashlyn (b.1988)	**Kanene** Marvin (b.1989)
Cosette Katie (b.1974) Lily (b.1977)	**Sue** Evan (b.1984)	**Katie** Louis (b.1990)
Lindsey Charlie (b.1977)	**Dawn** Shawn (b.1985) Brittany (b.1988)	**Debra** Justin (b.1990)

Appendix C: Race/Culture and Social Class of Participants

Participant	*Race/Culture*	*Social Class*
Della	White	middle
Cam	White	upper
Mimi	White	upper
Lindsey	White	upper middle
Cosette	White	upper
Rosemary	White	upper middle
Marie	White	upper middle
Chloe	White	upper middle
Sue	White	upper middle
Dawn	White	lower middle
Elsie	Hispanic	at or below poverty
Kim	White/Appalachian	middle
Kanene	African American	at or below poverty
Katie	White	middle
Debra	White/Italian American	middle

Appendix D: Group Interview Participants

Participants	*Generation*	*Race/Culture*	*Social Class*
Mimi	The Early Years	White	upper
Lindsey	The Early Years	White	upper middle
Sue	The Implementation Years	White	upper middle
Kim	The Maintenance Years	White/Appalachian	middle
Katie	The Maintenance Years	White	middle

Appendix E: Timeline

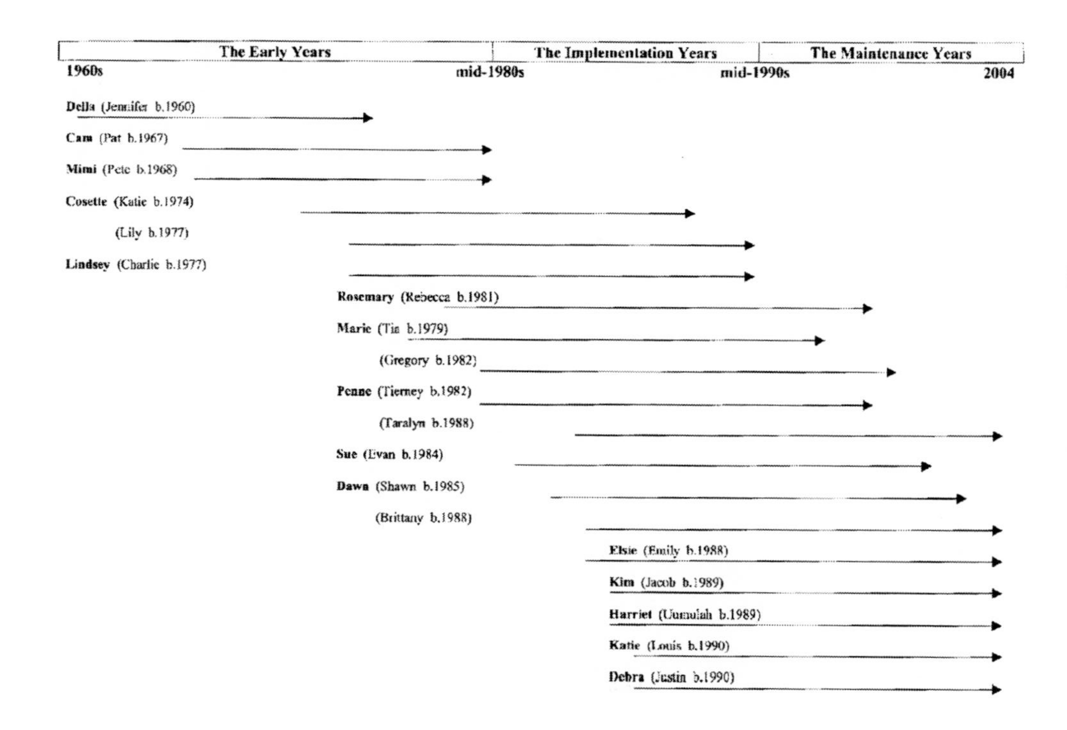

Appendix F: Katie's Drawing

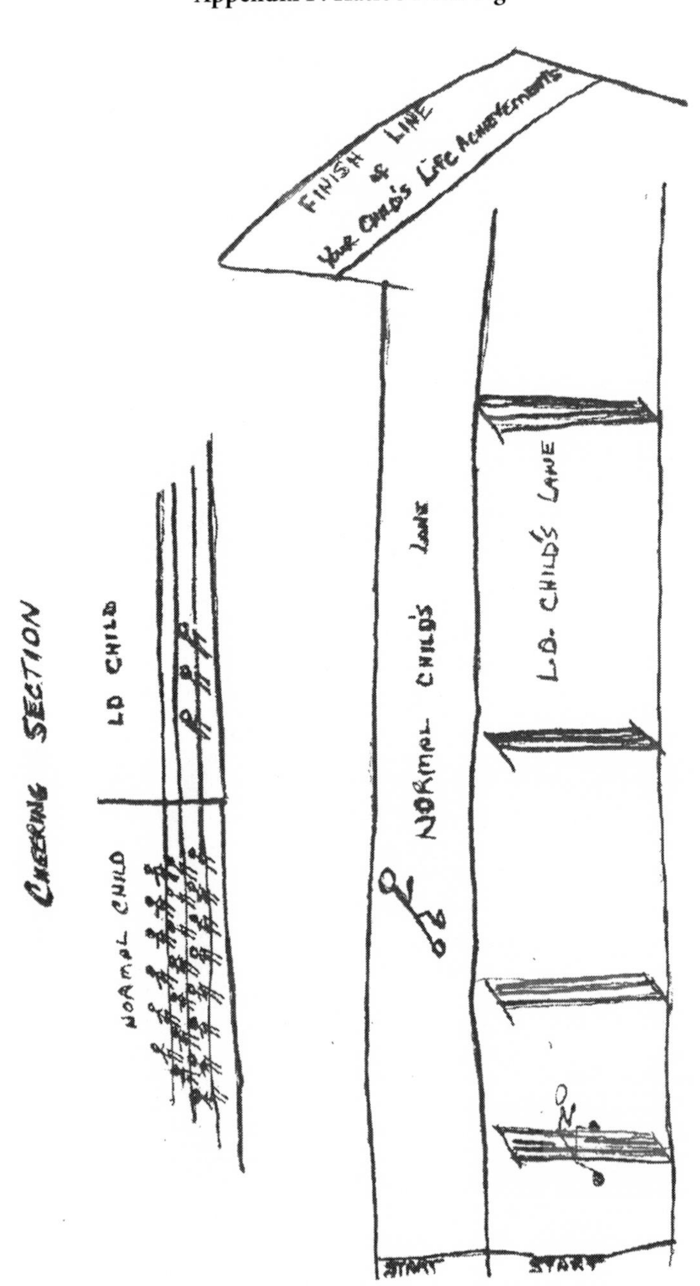

Appendix G: Kim's Drawing

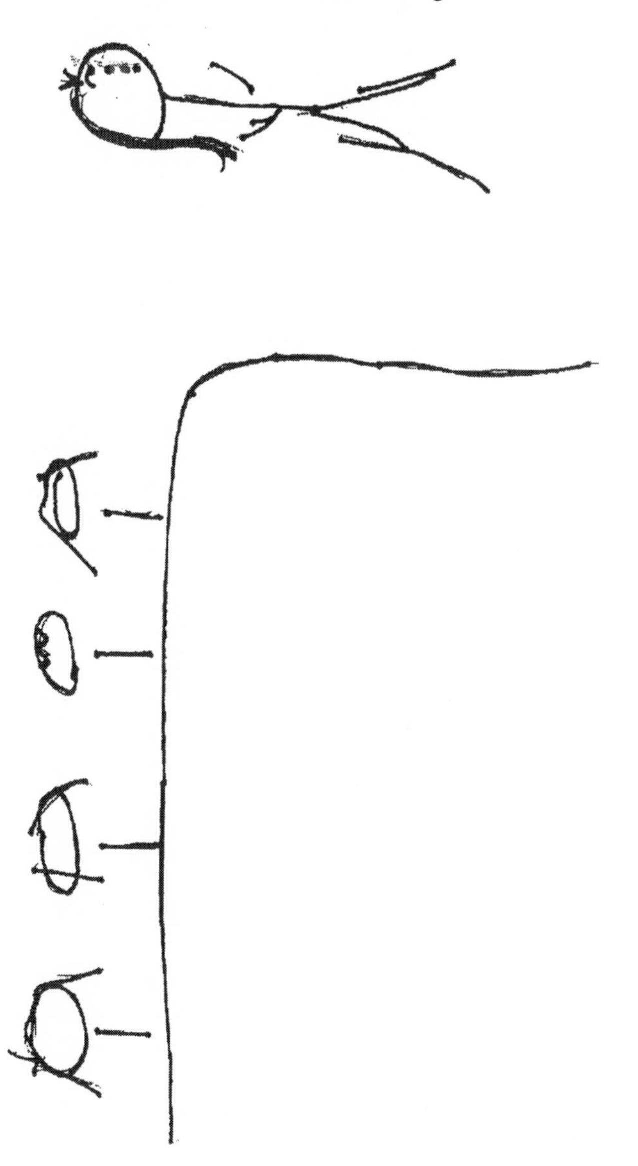

Appendix H: Lindsey's Drawing

Appendix I: Sue's Drawing

Appendix J: Mimi's Drawing

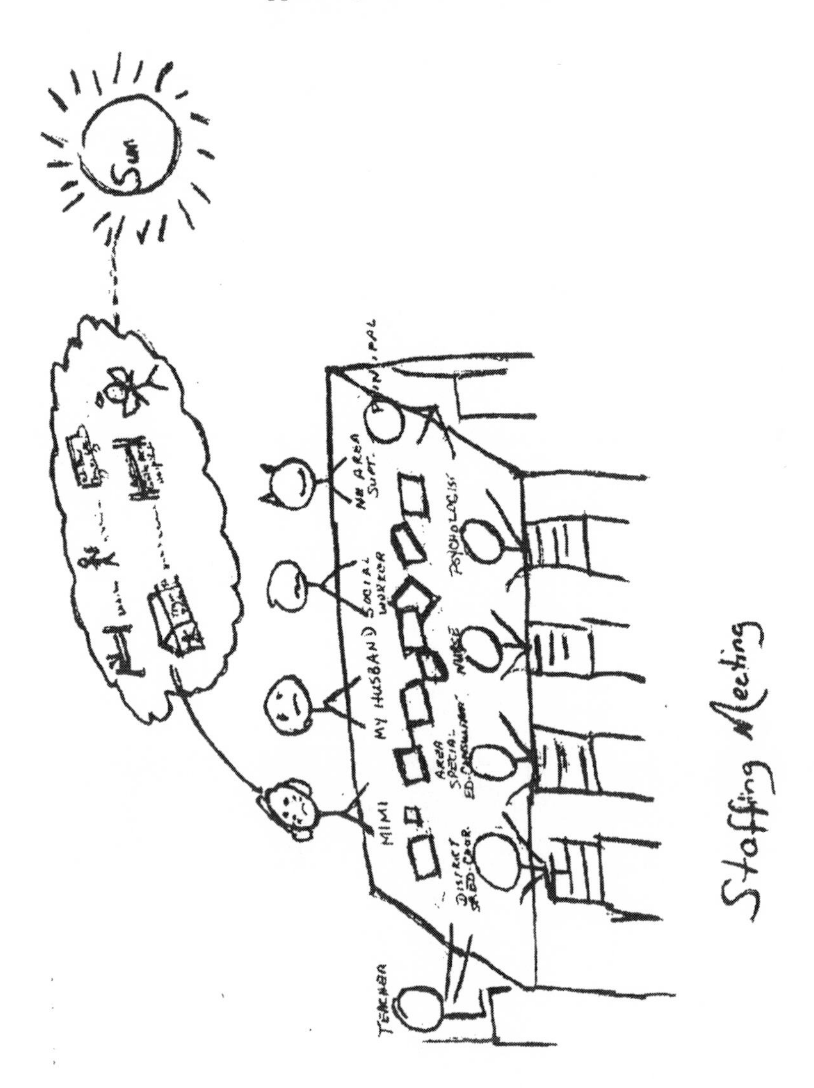

References

A Nation at Risk: The Imperative for Educational Reform. Washington, DC: The Commission on Excellence in Education, 1983.

Allan, J. (1999). *Actively seeking inclusion: Pupils with special needs in mainstream schools.* London: Falmer.

Anderson, J. D. (1988). *The education of Blacks in the south, 1860–1935.* Chapel Hill, NC: University of North Carolina Press.

Anselmo, S. S. (1977). Parent involvement in schools. *Clearing House, 50* (7), 297–299.

Apple, M. W. (1992). Education, culture, and class power: Basil Bernstein and the neo-Marxist sociology of education. *Education Theory, 42* (2), 127–145.

Apple, M. W. (1993). Foreword. In Casey, K., *I answer with my life: Life histories of women teachers working for social change* (pp. i–xii). New York: Routledge.

Artiles, A. J. (1998). The dilemma of difference: Enriching the disproportionality discourse with theory and context. *Journal of Special Education, 32,* 32–36.

Artiles, A. J., & Trent, S. (1994). Overrepresentation of minority students in special education: A continuing debate. *Journal of Special Education, 27,* 410–437.

Bakhtin, M. M. (1981). *The dialogic imagination: Four essays by M. M. Bakhtin.* Austin, TX: University of Texas Press.

Bakhtin, M. M. (1986). *Speech genres and other late essays.* Austin, TX: University of Texas Press.

Ballard, K., & McDonald, T. (1999). "Childhood" in the emergence and spread of U.S. public schooling. In T. Popkewitz & M. Brennan (Eds.), *M. Foucault's challenge: Discourse, power and knowledge* (pp. 117–143). New York: Teachers College Press.

Banks. J. (1991). *Teaching strategies for ethnic studies.* Boston, MA: Allyn & Bacon.

Barton, L. (2004). The politics of special education: A necessary or irrelevant approach? In L. Ware (Ed.), *Ideology and the politics of (in)exclusion* (pp. 63–75).

Barton, L., & Oliver, M. (1997). Special needs: Personal trouble or public issue? In B. Cosin & M. Hales (Eds.), *Families, education and social difference* (pp. 89–101). London & New York: Routledge in association with the Open University.

Baumberger, J. P., & Harper, R. E. (1999). *Assisting students with disabilities: What school counselors can and must do.* Thousand Oaks, CA: Corwin Press.

Bean, G., & Thorburn, M. (1995). Mobilising parent of children with disabilities in Jamaica and the English speaking Caribbean. In B. O'Toole & R. McConkey (Eds.), *Innovations in developing countries for people with disabilities* (pp. 105–120). Chorley, UK: Lisieux Hall in association with Associazione Italiana Amici di Raoul Follereau.

Belenky, M. F., Clinchy, B. M., Goldberger, N. R., & Tarule, J. M. (1986). *Women's ways of knowing: Development of self, voice, and mind.* New York: Basic Books.

Belenky, M. F., Clinchy, B. M., Goldberger, N. R., & Tarule, J. M. (1997). Toward an education for women. In D .J. Flinders & S. J. Thornton (Eds.), *The curriculum studies reader* (pp. 306–323). New York: Routledge.

Bell, D. A. (1978). The community role in the education of poor black children. *Theory into Practice, 17* (2), 115–121.

Bell, D. A. (1994). *Confronting authority: Reflections of an ardent protester.* Boston, MA: Beacon Press.

Bettleheim, B. (1976). *The uses of enchantment: The meaning and importance of fairy tales.* New York: Random House.

Blatt, B., & Kaplan, F. (1966). *Christmas in purgatory: A photographic essay on mental retardation.* Syracuse, NY: Human Policy Press.

Bloom, L. R. (1992). "How can we know the dancer from the dance?": Discourses of the self-body. *Human Studies, 15*, 313–334.

Bogdan, R. C., & Biklen, S. K. (1998). *Qualitative research for education: An introduction to theory and methods* (3rd ed.). Needham Heights, MA: A Viacom Company.

Brantlinger, E. A. (1996). The influence of preservice teachers' beliefs about pupil achievement on attitudes toward inclusion. *Teacher Education and Special Education, 19* (1), 17–33.

Brantlinger, E. A. (1997). Using ideology: Cases of non-recognition of the politics of research and practice in special education. *Review of Educational Research, 67*, 425–460.

Brantlinger, E. A. (2003). *Dividing classes: How the middle class negotiates and rationalizes school advantage.* New York: Routledge/Falmer.

Brantlinger, E. A. (2004). Confounding the needs and confronting the norms: An extension of Reid and Valle's essay. *Journal of Learning Disabilities, 37* (6), 490–499.

Brantlinger, E. A., Majd-Jabbari, M., & Guskin, S. (1996). Self-interest and liberal educational discourse: How ideology works for middle-class mothers. *American Educational Research Journal, 33* (3), 571–598.

Brueggemann, B. J. (1999). *Lend me your ear.* Washington, DC: Gallaudet University Press.

Bruner, J. S. (1986). *Actual minds, possible worlds.* Cambridge, MA: Harvard University Press.

Bruner, J. S. (1990). *Acts of meaning.* Cambridge, MA: Harvard University Press.

Campbell, J. (1949). *The hero with a thousand faces.* New York: Pantheon.

Casey, K. (1993). *I answer with my life: Life histories of women teachers working for social change.* New York: Routledge.

Clark, K., & Holquist, M. (1984). *Mikhail Bakhtin.* Cambridge, MA: Harvard University Press.

Code, L. (1991). *What can she know.* Ithaca, NY: Cornell University Press.

Conant, M. (1971). Teachers and parents changing roles and goals. *Childhood Education, 48* (3), 114–118.

Conley, D. (1999). *Being black, living in the red: Race, wealth, and social policy in America.* Berkeley, CA: University of California Press.
Connell, R. (1998). Masculinities and globalization. *Men and Masculinities, 1* (1), 2–23.
Connor, D. (2007). *Urban narratives: Portraits in Progress.* New York: Peter Lang.
Connor, M. H., & Boskin, J. (2001). Overrepresentation of bilingual and poor children in special education classes: A continuing problem. *Journal of Children & Poverty, 7,* 23–32.
Corker, M., & Shakespeare, T. (2002). *Disability/postmodernity: Embodying disability theory.* New York: Continuum.
Crenshaw, K. W. (1993). Beyond racism and misogyny: Black feminism and 2 Live Crew. In M. J. Matsuda, C. R. Lawrence, R. Delgado, & K. W. Crenshaw (Eds.), *Words that wound: Critical race theory, assaultive speech, and the first amendment* (pp. 111–132). Boulder, CO: Westview Press.
Cuban, L. (1993). *How teachers taught: Constancy and change in American classrooms, 1890–1990* (2nd ed.). New York: Teachers College Press.
Cutler, W. W. III. (2000). *Parents and schools: The 150-year struggle for control in American education.* Chicago, IL: University of Chicago Press.
Darder, A. (1991). *Culture and power in the classroom: A critical foundation for bicultural education.* Westport, CT: Bergin & Garvey.
Darder, A., Baltodano, M., & Torres, R. (2003). *The critical pedagogy reader.* New York: Routledge/Falmer.
Davis, L. J. (1997). Constructing normalcy. In L. J. Davis (Ed.), *Disability studies reader* (pp. 9–28). New York: Routledge.
Delgado, R. (1990). When a story is just a story: Does voice really matter? *Virginia Law Review, 76,* 95–111.
Delpit, L. (1995). *Other people's children: Cultural conflict in the classroom.* New York: New Press.
Dreyfus, H. L., & Rabinow, P. (1983). *Michel Foucault: Beyond structuralism and hermeneutics.* Chicago, IL: University of Chicago Press.
Education for All Handicapped Children Act (PL 94–142) 1975, amending Education of the Handicapped Act, renamed Individuals with Disabilities Education Act, as amended by PL 98–199, PL 99–457, PL 100–630, & PL 100–476, 20 U.S.C., Secs. 1400–1485.
Egan, K. (1991). Primary understanding: Education in early childhood. New York: Routledge.
Engel, D. M. (1993). Origin myths: Narratives of authority, resistance, disability, and law. *Law & Society Review, 27* (4), 785–826.
Evans, W. J., & Perry, C. (1991). *The impact of school-based management onschool environment.* Paper presented at the Annual Meeting of the American Educational Research Association, Chicago, April, p. 5.
Fadiman, A. (1997). *The spirit catches you and you fall down: A Hmong child, her American doctors, and the collision of two cultures.* New York: Farrar, Straus, & Giroux.
Fairclough, N. (1989). *Language and power.* London: Longman.
Fairclough, N. (1995). *Critical discourse analysis: The critical study of language.* London: Longman.

Feldman, L. (1989). Low-income parents becoming professionals with children—USA. *Early Child Development and Care, 50,* 151–158.

Ferri, B. A., & Connor, D. J. (2006). *Reading resistance: Discourses of exclusion in desegregation and inclusion debates.* New York: Peter Lang.

Finney, R. S. (1928). *A sociological philosophy of education.* New York: Macmillan.

Foster, M. (1997). *Black teachers on teaching.* New York: New Press.

Foucault, M. (1965). *Madness and civilization: A history of insanity in the age of reason.* (R. Howard, Trans.). New York: Vintage Books.

Foucault, M. (1972). *The archaeology of knowledge & the discourse on language.* (A. Smith, Trans.). New York: Pantheon Books.

Foucault, M. (1973). *The birth of the clinic: An archaeology of medical perception.* (A. Smith, Trans.). New York: Vintage Books.

Foucault, M. (1975). *Discipline & punish: The birth of the prison* (A. Sheridan, Trans.). New York: Vintage Books.

Foucault, M. (1977). *Language, counter-memory, practice: Selected essays and interviews* (D. Bouchard & S. Simon, Trans.). Ithaca, NY: Cornell University Press.

Foucault, M. (1978). *The history of sexuality* (R. Hurley, Trans.). New York: Vintage Books.

Foucault, M. (1980). *Power/knowledge: Selected interviews & other writings 1972–1977 by Michel Foucault* (C. Gordon, Ed.; G. Gordon, L. Marshall, J. Mepham, & K. Soper, Trans.). New York: Pantheon Books.

Foucault, M. (1983). Afterword: The subject and power. In Dreyfus, H. L., & Rabinow, P., *Michel Foucault: Beyond structuralism and hermeneutics.* Chicago, IL: University of Chicago Press.

Frey, J. H., & Fontana, A. (1993). The group interview in social research. In D. L. Morgan (Ed.), *Successful focus groups* (pp. 20–34). Newbury Park, CA: Sage.

Friere, P. (1997). Pedagogy of the oppressed. In D. Flinders & S. Thornton (Eds.), *The curriculum studies reader* (pp. 150–157). New York: Routledge.

Gee, J. P. (1999). *An introduction to discourse analysis: Theory and method.* London: Routledge.

Geertz, C. (1973). *The interpretation of cultures.* New York: Basic Books.

Gerschick, T. J. (1998). Sisyphus in a wheelchair: Men with physical disabilities confront gender domination. In J. O'Brien & J. Howard (Eds.), *Everyday inequalities: Critical inquiries* (pp. 189–211). Malden, MA: Blackwell.

Giorgi, A. (1994). A phenomenological perspective on certain qualitative research methods. *Journal of Phenomenological Psychology, 25,* 190–220.

Goldstein, S., Strickland, B., Turnbull, A. P., & Curry, L. (1980). An observational analysis of the IEP conference. *Exceptional Children, 46* (4), 278–286.

Gordon, C. (1980). Afterword. In C. Gordon (Ed.), *Power/knowledge: Selected interviews and other writings 1972–1977 by Michel Foucault* (pp. 229–259). New York: Pantheon Books.

Gould, S. J. (1996). *The mismeasure of man.* New York: W.W. Norton.

Greene, M. (1978). *Landscapes of learning.* New York: Teachers College Press.

Greene, M. (1995). *Releasing the imagination: Essays on education, the arts, and social change.* San Francisco: Jossey-Bass.

Greene, M. (1998). *The passionate mind of Maxine Greene* (W. F. Pinar, Ed.). Bristol, PA: Falmer. Taylor & Francis.

Grosz, E. (1993). Bodies and knowledges: Feminism and the crisis of reason. In E. Potter & L. Alcoff (Eds.), *Feminist epistemologies.* New York: Routledge.
Grumet, M. R., & Stone, L. (2000). Feminism and curriculum: Getting our act together. *Journal of Curriculum Studies, 32* (2), 183–197.
Habermas, J. (1978). *Knowledge and human interest* (2nd ed., J. Shapiro, Trans.). London: Heinemann.
Hall, E. T. (1981). *Beyond culture.* Garden City, NJ: Anchor Press/Doubleday.
Hallahan, D. P., & Cruickshank, W. M. (1973). *Psychoeducational foundations of learning disabilities.* Englewood Cliffs, NJ: Prentice-Hall.
Hanson, F. A. (1993). *Testing, testing: Social consequences of the examined life.* Berkeley, CA: University of California Press.
Haraway, D. (1988). Situated knowledges: The science question in feminism and the privilege of partial perspective. *Feminist Studies, 14,* 575–599.
Harding, S. (1998). *Multiculturalism, post-colonialism, and science: Epistemological issues.* Paper presented at the Annual American Educational Research Association Conference, San Diego.
Harre, R. (1990). Some narrative conventions of scientific discourse. In C. Nash (Ed.), *Narrative in culture: The uses of storytelling in the sciences, philosophy, and literature* (pp. 81–101). London: Routledge.
Harry, B. (1992). An ethnographic study of cross-cultural communication with Puerto Rican-American families in the special education system. *American Educational Research Journal 29* (3), 471–494.
Harry, B. (1994). *The disproportionate representation of minority students in special education: Theories and recommendations.* Alexandria, VA: National Association of State Directors of Special Education.
Harry, B., Allen, N., & McLaughlin, M. (1995). Communication versus compliance: African-American parents' involvement in special education. *Exceptional Children, 61* (4), 364–377.
Hartocollis, A. (1998, November). U.S. questions the placement of city pupils. *New York Times.*
Haug et al. (1987). *Female sexualization: A collective work of memory.* New York: Verso.
Heleen, O. (1992). Schools reaching out: An introduction. *Equity and Choice, 8* (2), 5–8.
Heshusius, L. (1994). Freeing ourselves from objectivity: Managing subjectivity or turning toward a participatory mode of consciousness? *Educational Researcher, 23,* 15–22.
Hirsch, E. D., Kett, J. F., & Trefil, J. (2002). *The new dictionary of cultural literacy: What every American needs to know.* Boston, MA: Houghton Mifflin.
Hoel, T. L. (1997). Voices from the classroom. *Teaching and Teacher Education, 13,* 5–16.
Hoff, M. K., Fenton, K. S., Yoshida, R. K., & Kaufman, M. J. (1978). Notice and consent: The school's responsibility to inform parents. *Journal of School Psychology, 16* (3), 265–272.
Holdsworth, N. (1995, May 19). "Culture of dignity" for special needs. *Times Educational Supplement,* p. 14.
hooks, b. (1989). *Talking back thinking feminist thinking black.* Boston, MA: South End.

Horn, G. (1970). Home visits. *Today's Education, 59* (6), 44–46.

Individuals with Disabilities Education Act (IDEA) of 1990, PL 101–476 20 U.S.C. Sec. 1400 et seq.

Individuals with Disabilities Education Act (IDEA). Amendments of 1997.

Individuals with Disabilities Education Act (IDEA). Amendments of 2004.

Joynson, R. B. (1974). *Psychology and common sense.* London: Routledge and Kegan Paul.

Kalyanpur, M., & Harry, B. (1999). *Culture in special education: Building reciprocal Family-professional relationships.* Baltimore, MD: Paul H. Brookes.

Kalyanpur, M., Harry, B., & Skrtic, T. (2000). Equity and advocacy expectations of culturally diverse families' participation in special education. *International Journal of Disability, Development, and Education, 47* (2), 119–136.

Kendall, G., & Wickham, G. (1999). *Using Foucault's methods.* Thousand Oaks, CA: Sage.

Kierkegaard, S. (1975). The first existentialist. In W. Kaufmann (Ed.), *Existentialism from Dostoevsky to Sartre* (pp. 83–120). New York: Penguin Books.

Kinchloe, J. L., & McLaren, P. L. (1998). Rethinking critical theory and qualitative research. In N. K. Denzin & Y. S. Lincoln (Eds.), *The landscape of qualitative research* (pp. 260–299). Thousand Oaks, CA: Sage.

Kinchloe, J. L., Steinberg, S. R., & Gresson, A. D. (1997). *Measured lies: The bell curve examined.* New York: Palgrave Macmillan.

Kinchloe, J. L., Steinberg, S. R., & Villaverde, L. E. (1999). *Rethinking intelligence: Confronting psychological assumptions about teaching and learning.* New York: Routledge.

Kirk, S. A. (1963). Behavioral diagnosis and remediation of learning disabilities. In *Proceedings of the Annual Meeting of the Conference on Exploration into the Problems of the Perceptually Handicapped Child (Vol. I).* Chicago, IL.

Kliebard, H. M. (1995). *The struggle for the American curriculum, 1893–1958* (2nd ed.). New York: Routledge.

Krueger, R. A. (1993). Quality control in focus group research. In D. L. Morgan (Ed.), *Successful focus groups* (pp. 65–88). Newbury Park, CA: Sage.

Kvale, S. (1996). *InterViews: An introduction to qualitative research interviewing.* Thousand Oaks, CA: Sage.

Labov, W., & Weletzky, J. (1967). Narrative analysis: Oral versions of personal experiences. In J. Helms (Ed.), *Essays on verbal and visual arts* (pp. 12–44). Seattle, WA: University of Washington Press.

Ladson-Billings, G., & Tate, W. (1995). Towards a critical race theory of education.*Teachers College Record, 97,* 47–68.

Lagrander, L. M., & Reid, D. K. (2000). Language acquisition and usage: Multicultural and multilinguistic perspectives. In K. Fahey & D. K. Reid (Eds.), *Language development, differences, and disorders* (pp. 177–218). Austin, TX: Pro-Ed.

Lerner, G. (1986). *The creation of patriarchy.* New York and Oxford: Oxford University Press.

Lincoln, Y. S. (1997). Self, subject, audience, text: Living at the edge, writing in the margins. In W. Tierney & Y. Lincoln (Eds.), *Representation and the text* (pp. 37–55). Albany, NY: State University of New York Press.

Linton, S. (1998). *Claiming disability.* New York: New York University Press.
Lipsky, D. K. (1985). A parental perspective on stress and coping. *American Journal of Orthopsychiatry, 55* (4), 614–617.
Lipsky, D. K., & Gartner, A. (1996). Inclusion, restructuring and the remaking of American society. *Harvard Educational Review, 66,* 762–796.
Lipsky, D. K., & Gartner, A. (1997). *Inclusion and school reform: Transforming America's classrooms.* Baltimore, MD: Paul H. Brookes.
Lomawaima, K. T. (1995). Domesticity in the federal Indian schools. The power of authority over mind. In J. Terry & J. Urla (Eds.), *Deviant bodies. Critical perspectives on differences in science and popular culture* (pp. 197–218). Bloomington, IN: Indiana University Press.
Lorde, A. (1984). *Sister outsider: Essays and speeches.* Trumansburg, NY: Crossing Press.
Losen, D. J., & Orfield, G. (2002). Racial inequity in special education. In D. J. Losen & G. Orfield (Eds.), *Racial inequity in special education* (pp. xv–xxxvii). Cambridge, MA: Harvard Education Press.
Lyotard, J. F. (1979). *The postmodern condition: A report on knowledge.* Minneapolis, MN: University of Minnesota Press.
Malekoff, A., Johnson, H., & Klappersack, B. (1991). Parent-professional collaboration on behalf of children with learning disabilities. *Families in Society: The Journal of Contemporary Human Services, 72,* 416–424.
Manicom, A. (1984). Feminist frameworks and teacher education. *Journal of Education, 166* (1), 77–88.
Martin, J. R. (1982). Excluding women from the educational realm. *Harvard Educational Review, 52* (2), 133–148.
McEwan, H. (1997). The functions of narrative and research on teaching. *Teaching and Teacher Education, 15* (1), 85–92.
McFall, M. L. (1974). Dear parents: The single parent and child care needs. *Contemporary Education, 45* (4), 289–291.
Mehan, H., Hartwick, A., & Meihls, J. L. (1996). *Handicapping the handicapped.* Stanford, CA: Stanford University Press.
Merleau-Ponty, M. (1964). *The primacy of perception* (J. M. Edie, Trans.). Evanston, IL: Northwestern University Press.
Meyers, P., & Hammill, D. (1990). *Learning disabilities: Basic concepts, assessment practices, and instructional strategies.* Austin, TX: Pro-Ed.
Mishler, E. G. (1999). *Craftartists' narratives of identity.* Cambridge, MA: Harvard University Press.
Monk, R. (1990). *Ludwig Wittgenstein: The duty of genius.* New York: Penguin Books.
Morgan, D. L., & Krueger, R. A. (1993). When to use focus groups and why. In D. L. Morgan (Ed.), *Successful focus groups* (pp. 3–19). Newbury Park, CA: Sage.
National Commission on Excellence in Education. (1983). *A nation at risk: The imperative for educational reform.* Washington, DC: U.S. Department of Education.
National Council on Disability. (1995). *Improving implementation of the Individuals with Disabilities Education Act: Making schools work for all of America's children.* Washington, DC: Author.

Nieto, S. (1995). From brown heroes and holidays to assimilationist agendas: Reconsidering the critiques of multicultural education. In C. E. Sleeter & P. L. McLaren (Eds.), *Multicultural education, critical pedagogy, and the politics of difference* (pp. 191–220). Albany, NY: SUNY Press.

Nieto, S. (1999). *The light in their eyes: Creating multicultural learning communities.* New York: Teachers College Press.

No Child Left Behind Act of 2001, PL 107–110 20 U.S.C. 6301 et seq.

Oakes, J., Gamoran, A., & Page, R. N. (1992). Curriculum differentiation: Opportunities, outcomes, and meanings. In P. W. Jackson (Ed.), *Handbook of research on curriculum* (pp. 570–608). Washington, DC: American Educational Research Association.

Olesen, V. (1998). Feminisms and models of qualitative research. In N. K. Denzin and Y. S. Lincoln (Eds.), *The landscape of qualitative research* (pp. 300–332). Thousand Oaks, CA: Sage.

Oliver, M. L., & Shapiro, T. M. (1995). *Black wealth/white wealth.* New York: Routledge.

Olmstead, P. P., & Jester, R. E. (1972). Mother-child interaction in a teaching situation. *Theory into Practice, 11* (3), 163–170.

Orton, S. A. (1937). *Reading, writing, and speech problems in children.* New York: W.W. Norton.

Padilla, A. M., & Lindholm, K. J. (1996). In W. F. Tate, IV, Critical race theory and education: History, theory, and implications. *Review of Research in Education, 22,* 195–247.

Poland, S. F., Thurlow, M. L., Ysseldyke, J. E., & Mirkin, P. K. (1982). Current psychoeducational assessment and decision-making practices as reported by directors of special education. *Journal of School Psychology, 20* (3), 171–178.

Polkinghorne, D. (1988). *Narrative knowing and the human sciences.* Albany, NY: State University of New York Press.

Poplin, M. S. (1988). The reductionist fallacy in learning disabilities: Replicating the past by reproducing the present. *Journal of Learning Disabilities, 21,* 389–400.

Price, J., & Shildrick, M. (1999). Breaking the boundaries of the broken body. In J. Price & M. Shildrick (Eds.), *Feminist theory and the body* (pp. 432–444). London: Taylor & Francis.

Reese, W. (1978). Between home and school: Organized parents, clubwomen, and urban education in the progressive era. *School Review, 8* (1), 3–28.

Reid, D. K., & Button, L. J. (1995). Anna's story: Narratives of personal experience about being labeled learning disabled. *Journal of Learning Disabilities, 28* (10), 602–614.

Reid, D. K., Hresko, W. P., & Swanson, H. L. (1996). *Cognitive approaches to learning Disabilities* (3rd ed.). Austin, TX: Pro-Ed.

Reid, D. K., & Valle, J. W. (2004). The discursive practice of learning disability: Implications for instruction and parent-school relations. *Journal of Learning Disabilities, 37* (6), 466–481.

Rickover, H. G. (1959). *Education and freedom.* New York: E. P. Dutton.

Rockowitz, R. J., & Davidson, P. W. (1979). Discussing diagnostic findings with parents. *Journal of Learning Disabilities, 12* (1), 11–16.

Rodis, P., Garrod, A., & Boscardin, M. L. (Eds.). (2001). *Learning disabilities and life stories.* Needham Heights, MA: Allyn & Bacon.

Rorty, R. (1989). *Contingency, irony, and solidarity.* Cambridge: Cambridge University Press.

Rousmaniere, K. (1997). *City teachers: Teaching and school reform in historical perspective.* New York: Teachers College Press.

Sapon-Shevin, M. (1999). *Because we can change the world: A practical guide to building cooperative, inclusive classroom communities.* Boston, MA: Allyn & Bacon.

Schutz, A. (1967). *The phenomenology of the social world.* Evanston, IL: Northwestern University Press.

Shiffrin, D. (1996). Narrative as self-portrait: Sociolinguistic constructions of identity. *Language in Society, 25,* 167–203.

Shildrick, M. (1997). *Leaky bodies and boundaries.* New York: Routledge.

Shildrick, M., & Price, J. (1999). *Vital signs: Feminist reconfigurations of the biological bodies.* Edinburgh, Scotland: Edinburgh University Press.

Skrtic, T. M. (1991). *Behind special education: A critical analysis of professional culture and school organization.* Denver, CO: Love.

Skrtic, T. M. (1995). *Disability & democracy: Reconstructing (special) education for postmodernity.* New York: Teachers College Press.

Sleeter, C. E. (1986). Learning disabilities: The social construction of a special education category. *Exceptional Children, 53* (1), 46–54.

Sleeter, C. E. (1987). Why is there learning disabilities? A critical analysis of the birth of the field in social context. In T. Popkewitz (Ed.), *The foundations of school subjects* (pp. 210–237). London: Falmer.

Smith, B. O. (1942). The war and the educational program. *Curriculum Journal, 13,* 113–116.

Sonnenschein, P. (1981). Parents and professionals: An uneasy relationship. *Teaching Exceptional Children, 14,* 62–65.

Spring, J. H. (1989). *The sorting machine revisited: National educational policy since 1945.* New York: Longman.

Spring, J. H. (2002). *Conflict of interests: The politics of American education* (4th ed.). New York: McGraw-Hill Higher Education.

Sternberg, L., Taylor, R. L., & Russell, S. C. (1996). *Negotiating the disability maze.* Springfield, IL: Charles C. Thomas.

Tate, W. F. IV. (1996). Critical race theory and education: History, theory, and implications. *Review of Research in Education, 22,* 195–247.

Thomas, G., & Loxley, A. (2001). *Deconstructing special education and constructing inclusion.* Philadelphia, PA: Open University Press.

Tooley, J. (1999). Asking different questions: Towards justifying markers in education. In N. Alexiadou & C. Brock (Eds.), *Education as a commodity* (pp. 9–19). Suffolk, England: John Catt Educational.

Traub, J. (2000, January 16). What no school can do. *New York Times Magazine* [Online], pp. 1–12. Available: ProQuest.

Turnbull, A. P., & Turnbull, H. R. (1997). *Families, professionals, and exceptionality: A special partnership* (3rd ed.). Upper Saddle River, NJ: Prentice-Hall. Simon & Schuster/A Viacom Company.

Tyack, D. J. (1974). *The one best system: A history of American urban education.* Cambridge, MA: Harvard University Press.

Tyack, D. J., & Tobin, W. (1994). The "grammar" of schooling: Why has it been so hard to change? *American Educational Research Journal, 31*, 453–479.

U.S. Bureau of the Census. (September, 1993). *Monthly News.*

Valle, J. W., & Aponte, E. (2002). IDEA: A Bakhtinian perspective on parent and professional discourse. *Journal of Learning Disabilities, 35* (5), 469–479.

Valle, J. W., & Reid, D. K. (2001). Parent-educator partnerships: A critical history of the search for authentic and respectful home-school relationships. *School Public Relations, 22* (1), 23–36.

Varenne, H., & McDermott, R. (1998). *Successful failure: The school America builds.* Boulder, CO: Westview Press.

Vice, S. (1997). *Introducing Bakhtin.* Manchester: Manchester University Press.

Volosinov, V. N. (1973). *Marxism and the philosophy of language* (L. Matejka & I. R. Titunik, Trans.). Cambridge, MA: Harvard University Press.

Ware, L. (2004). *Ideology and the politics of (in)exclusion.* New York: Peter Lang.

Warner, J. (2005). *Perfect madness: Motherhood in the age of anxiety.* New York: Riverhead Books.

Wertsch, J. V. (1991). *Voices of the mind: A sociocultural approach to mediated action.* Cambridge, MA: Harvard University Press.

Wolfensberger, W., & Nirje, B. (1972). *The principle of normalization in human services.* Toronto: National Institute on Mental Retardation.

Young, I. M. (1990). *Throwing like a girl and other essays in feminist philosophy and social theory.* Bloomington, IN: Indiana University Press.

Index

LaVergne, TN USA
29 June 2010
187755LV00001B/44/P